Math Ma   # 2
Nice-n-E~~y~~ Books™

In This Book:
- ## Percentages
- ## Exponents
- ## Radicals
- ## Logarithms
- ## Algebra Basics

"MATH MADE NICE-n-EASY #2" is one in a series of books designed to make the learning of math interesting and fun. For help with additional math topics, see the complete series of "MATH MADE NICE-n-EASY" titles.

Based on U.S. Government
Teaching Materials

Research & Education Association
61 Ethel Road West
Piscataway, New Jersey 08854

MATH MADE NICE-N-EASY BOOKS™
BOOK #2

Year 2003 Printing

Printed in the United States of America

Library of Congress Control Number 99-70141

International Standard Book Number 0-87891-201-0

MATH MADE NICE-N-EASY is a trademark of
Research & Education Association, Piscataway, New Jersey 08854

WHAT "MATH MADE NICE-N-EASY" WILL DO FOR YOU

The "Math Made Nice-n-Easy" series simplifies the learning and use of math and lets you see that math is actually interesting and fun. This series of books is for people who have found math scary, but who nevertheless need some understanding of math without having to deal with the complexities found in most math textbooks.

The "Math Made Nice-n-Easy" series of books is useful for students and everyone who needs to acquire a basic understanding of one or more math topics. For this purpose, the series is divided into a number of books which deal with math in an easy-to-follow sequence beginning with basic arithmetic, and extending through pre-algebra, algebra, and calculus. Each topic is described in a way that makes learning and understanding easy.

Almost everyone needs to know at least some math at work, or in a course of study.

For example, almost all college entrance tests and professional exams require solving math problems. Also, almost all occupations (waiters, sales clerks, office people) and all crafts (carpentry, plumbing, electrical) require some ability in math problem solving.

The "Math Made Nice-n-Easy" series helps the reader grasp quickly the fundamentals that are needed in using math. The reader is led by the hand, step-by-step, through the various concepts and how they are used.

By acquiring the ability to use math, the reader is encouraged to further his/her skills and to forget about any initial math fears.

The "Math Made Nice-n-Easy" series includes material originated by U.S. Government research and educational efforts. The research was aimed at devising tutoring and teaching methods for educating government personnel lacking a technical and/or mathematical background. Thanks for these efforts are due to the U.S. Bureau of Naval Personnel Training.

Dr. Max Fogiel
Program Director

Contents

Chapter 8
LOGARITHMS

Chapter 9
FUNDAMENTALS OF ALGEBRA

CHAPTER 6

PERCENTAGE AND MEASUREMENT

In the discussion of decimal fractions, it was shown that for convenience in writing fractions whose denominators are 10 or some power of 10, the decimal point could be employed and the denominators could be dropped. Thus, this special group of fractions could be written in a much simpler way. As early as the 15th century, businessmen made use of certain decimal fractions so much that they gave them the special designation PERCENT.

MEANING OF PERCENT

The word "percent" is derived from Latin. It was originally "per centum," which means "by the hundred." Thus the statement is often made that "percent means hundredths."

Percentage deals with the group of decimal fractions whose denominators are 100—that is, fractions of two decimal places. Since hundredths were used so frequently, the decimal point was dropped and the symbol % was placed after the number and read "percent" (per 100). Thus, 0.15 and 15% represent the same value,

15/100. The first is read "15 hundredths," and the second is read "15 percent." Both mean 15 parts out of 100.

Ordinarily, percent is used in discussing relative values. For example, 25 percent may convey an idea of relative value or relationship. To say "25 percent of the crew is ashore" gives an idea of what part of the crew is gone, but it does not tell how many. For example, 25 percent of the crew would represent vastly different numbers if the comparison were made between an LSM and a cruiser. When it is necessary to use a percent in computation, the number is written in its decimal form to avoid confusion.

By converting all decimal fractions so that they had the common denominator 100, men found that they could mentally visualize the relative size of the part of the whole that was being considered.

CHANGING DECIMALS TO PERCENT

Since percent means hundredths, any decimal may be changed to percent by first expressing it as a fraction with 100 as the denominator. The numerator of the fraction thus formed indicates how many hundredths we have, and therefore it indicates "how many percent" we have. For example, 0.36 is the same as 36/100. Therefore, 0.36 expressed as a percentage would be 36 percent. By the same reasoning, since 0.052 is equal to 5.2/100, 0.052 is the same as 5.2 percent.

In actual practice, the step in which the denominator 100 occurs is seldom written down. The expression in terms of hundredths is converted mentally to percent. This results in the following rule: To change a decimal to percent, multiply the decimal by 100 and annex the percent sign (%). Since multiplying by 100 has the effect of moving the decimal point two places to the right, the rule is sometimes stated as follows: To change a decimal to percent, move the decimal point two places to the right and annex the percent sign.

Changing Common Fractions and Whole Numbers to Percent

Common fractions are changed to percent by first expressing them as decimals. For example, the fraction 1/4 is equivalent to the decimal 0.25. Thus 1/4 is the same as 25 percent.

Whole numbers may be considered as special types of decimals (for example, 4 may be written as 4.00) and thus may be expressed in terms of percentage. The meaning of an expression such as 400 percent is vague unless we keep in mind that percentage is a form of comparison. For example, a question which often arises is "How can I have more than 100 percent of something, if 100 percent means all of it?"

This question seems reasonable, if we limit our attention to such quantities as test scores. However, it is also reasonable to use percentage in comparing a current set of data with a previous set. For example, if the amount of electrical power used by a Navy facility this year is double the amount used last year, then this year's power usage is 200 percent of last year's usage.

The meaning of a phrase such as "200 percent of last year's usage" is often misinterpreted. A total amount that is 200 percent of the previous amount is not the same as an increase of 200 percent. The increase in this case is only 100 percent, for a total of 200. If the increase had been 200 percent, then the new usage figure would be 300 percent of the previous figure.

Baseball batting averages comprise a special case in which percentage is used with only occasional reference to the word "percent." The percentages in batting averages are expressed in their decimal form, with the figure 1.000 representing 100 percent. Although a batting average of 0.300 is referred to as "batting 300," this is actually erroneous nomenclature from the strictly mathematical standpoint. The correct statement, mathematically, would be "batting point three zero zero" or "batting 30 percent."

Practice problems. Change each of the following numbers to percent:

1. 0.0065	3. 0.363	5. 7
2. 1.25	4. 3/4	6. 1/2

Answers:

1. 0.65%	3. 36.3%	5. 700%
2. 125%	4. 75%	6. 50%

CHANGING A PERCENT TO A DECIMAL

Since we do not compute with numbers in the

percent form, it is often necessary to change a percent back to the decimal form. The procedure is just opposite to that used in changing decimals to percents: To change a percent to a decimal, drop the percent sign and divide the number by 100. Mechanically, the decimal point is simply shifted two places to the left and the percent sign is dropped. For example, 25 percent is the same as the decimal 0.25. Percents larger than 100 percent are changed to decimals by the same procedure as ordinary percents. For example, 125 percent is equivalent to 1.25.

Practice problems. Change the following percents to decimals:

1. 2.5%   3. 125%   5. $5\frac{3}{4}\%$

2. 0.63%   4. 25%   6. $9\frac{1}{2}\%$

Answers:

1. 0.025   3. 1.25   5. 5.75% = 0.0575
2. 0.0063   4. 0.25   6. 9.50% = 0.095

THE THREE PERCENTAGE CASES

To explain the cases that arise in problems involving percents, it is necessary to define the terms that will be used. Rate (r) is the number of hundredths parts taken. This is the number followed by the percent sign. The base (b) is the whole on which the rate operates. Percentage (p) is the part of the base determined by the rate. In the example

$$5\% \text{ of } 40 = 2$$

5% is the rate, 40 is the base, and 2 is the percentage.

There are three cases that usually arise in dealing with percentage, as follows:

Case I—To find the percentage when the base and rate are known.

EXAMPLE: What number is 6% of 50?

Case II—To find the rate when the base and percentage are known.

EXAMPLE: 20 is what percent of 60?

Case III—To find the base when the percentage and rate are known.

EXAMPLE: The number 5 is 25% of what number?

Case I

In the example

$$6\% \text{ of } 50 = ?$$

the "of" has the same meaning as it does in fractional examples, such as

$$\frac{1}{4} \text{ of } 16 = ?$$

163

In other words, "of" means to multiply. Thus, to find the percentage, multiply the base by the rate. Of course the rate must be changed from a percent to a decimal before multiplying can be done. Rate times base equals percentage. Thus,

$$6\% \text{ of } 50 = ?$$

$$0.06 \times 50 = 3$$

The number that is 6% of 50 is 3.

FRACTIONAL PERCENTS.—A fractional percent represents a part of 1 percent. In a case such as this, it is sometimes easier to find 1 percent of the number and then find the fractional part. For example, we would find 1/4 percent of 840 as follows:

$$1\% \text{ of } 840 = 0.01 \times 840$$

$$= 8.40$$

Therefore, $\frac{1}{4}\%$ of $840 = 8.40 \times \frac{1}{4}$

$$= 2.10$$

Case II

To explain case II and case III, we notice in the foregoing example that the base corresponds to the multiplicand, the rate corresponds to the multiplier, and the percentage corresponds to the product.

$$50 \text{ (base or multiplicand)}$$
$$\underline{.06} \text{ (rate or multiplier)}$$
$$3.00 \text{ (percentage or product)}$$

Recalling that the product divided by one of its factors gives the other factor, we can solve the following problem:

$$?\% \text{ of } 60 = 20$$

We are given the base (60) and percentage (20).

60 (base)
  ? (rate)
————————
20 (percentage)

We then divide the product (percentage) by the multiplicand (base) to get the other factor (rate). Percentage divided by base equals rate. The rate is found as follows:

$$\frac{20}{60} = \frac{1}{3}$$

$$= .33\frac{1}{3}$$

$$= 33\frac{1}{3}\% \text{ (rate)}$$

The rule for case II, as illustrated in the foregoing problem, is as follows: To find the rate when the percentage and base are known, divide the percentage by the base. Write the quotient in the decimal form first, and finally as a percent.

Case III

The unknown factor in case III is the base, and the rate and percentage are known.

EXAMPLE: 25% of ? = 5

$$\begin{array}{r} \text{? (base)} \\ \underline{.25 \text{ (rate)}} \\ 5.00 \text{ (percentage)} \end{array}$$

We divide the product by its known factor to find the other factor. Percentage divided by rate equals base. Thus,

$$\frac{5}{.25} = 20 \text{ (base)}$$

The rule for case III may be stated as follows: To find the base when the rate and percentage are known, divide the percentage by the rate.

Practice problems. In each of the following problems, first determine which case is involved; then find the answer.

1. What is $\frac{3}{4}$% of 740?

2. 7.5% of 2.75 = ?

3. 8 is 2% of what number?

4. ?% of 18 = 15.

5. 12% of ? = 12.

6. 8 is what percent of 32?

Answers:

1. Case I; 5.55

2. Case I; 0.20625

3. Case III; 400

166

4. Case II; $83\frac{1}{3}\%$

5. Case III; 100

6. Case II; 25%

PRINCIPLES OF MEASUREMENT

Computation with decimals frequently involves the addition or subtraction of numbers which do not have the same number of decimal places. For example, we may be asked to add such numbers as 4.1 and 32.31582. How should they be added? Should zeros be annexed to 4.1 until it is of the same order as the other decimal (to the same number of places)? Or, should .31582 be rounded off to tenths? Would the sum be accurate to tenths or hundred-thousandths? The answers to these questions depend on how the numbers orignially arise.

Some decimals are finite or are considered as such because of their use. For instance, the decimal that represents $\frac{1}{2}$, that is 0.5, is as accurate at 0.5 as it is at 0.5000. Likewise, the decimal that represents $\frac{1}{8}$ has the value 0.125 and could be written just as accurately with additional end zeros. Such numbers are said to be finite. Counting numbers are finite. Dollars and cents are examples of finite values. Thus, $10.25 and $5.00 are finite values.

To add the decimals that represent $\frac{1}{8}$ and $\frac{1}{2}$, it is not necessary to round off 0.125 to tenths. Thus, 0.5 + 0.125 is added as follows:

167

$$0.500$$
$$\underline{0.125}$$
$$0.625$$

Notice that the end zeros were added to 0.5 to carry it out the same number of places as 0.125. It is not necessary to write such place-holding zeros if the figures are kept in the correct columns and decimal points are alined. Decimals that have a definite fixed value may be added or subtracted although they are of different order.

On the other hand, if the numbers result from measurement of some kind, then the question of how much to round off must be decided in terms of the precision and accuracy of the measurements.

ESTIMATION

Suppose that two numbers to be added resulted from measurement. Let us say that one number was measured with a ruler marked off in tenths of an inch and was found, to the nearest tenth of an inch, to be 2.3 inches. The other number measured with a precision rule was found, to the nearest thousandth of an inch, to be 1.426 inches.

Each of these measurements requires estimation between marks on the rule, and estimation between marks on any measuring instrument is subject to human error. Experience has shown that the best the average person can do with consistency is to decide whether a measurement is more or less than halfway between marks. The correct way to state this fact mathematically is to say that a measurement made with an instrument marked off in tenths of an inch involves a maximum probable error of 0.05 inch (five hundredths is one-half of one tenth). By the same reasoning, the probable error in a measurement made with an instrument marked in thousandths of an inch is 0.0005 inch.

PRECISION

In general, the probable error in any measurement is one-half the size of the smallest division on the measuring instrument. Thus the precision of a measurement depends upon how precisely the instrument is marked. It is important to realize that precision refers to the size of the smallest division on the scale; it has nothing to do with the accuracy (correctness) of the markings. In other words, to say that one instrument is more precise than another does not imply that the less precise instrument is poorly manufactured. In fact, it would be possible to make an instrument with very high apparent precision, and yet mark it carelessly so that measurements taken with it would be inaccurate.

From the mathematical standpoint, the precision of a number resulting from measurement depends upon the number of decimal places; that is, a larger number of decimal places means a smaller probable error. In 2.3 inches the probable error is 0.05 inch, since 2.3 actually lies somewhere between 2.25 and 2.35. In 1.426 inches there is a much smaller probable error of 0.0005 inch. If we add 2.300 + 1.426

and get an answer in thousandths, the answer, 3.726 inches, would appear to be precise to thousandths; but this is not true since there was a probable error of .05 in one of the addends. Also 2.300 appears to be precise to thousandths but in this example it is precise only to tenths. It is evident that the precision of a sum is no greater than the precision of the least precise addend. It can also be shown that the precision of a difference is no greater than the less precise number compared.

To add or subtract numbers of different orders, all numbers should first be rounded off to the order of the least precise number. In the foregoing example, 1.426 should be rounded to tenths—that is, 1.4.

This rule also applies to repeating decimals. Since it is possible to round off a repeating decimal at any desired point, the degree of precision desired should be determined and all repeating decimals to be added should be rounded to this level. Thus, to add the decimals generated by $\frac{1}{3}$, $\frac{2}{3}$, and $\frac{5}{12}$ correct to thousandths, first round off each decimal to thousandths, and then add, as follows:

$$
\begin{array}{r}
.333 \\
.667 \\
\underline{.417} \\
1.417
\end{array}
$$

When a common fraction is used in recording the results of measurement, the denominator of the fraction indicates the degree of precision. For example, a ruler marked in sixty-fourths of an inch has much smaller divisions than one marked in fourths of an inch. Therefore a measurement of $3\frac{4}{64}$ inches is more precise than a measure of $3\frac{1}{4}$ inches, even though the two fractions are numerically equal. Remember that a measurement of $3\frac{4}{64}$ inches contains a probable error of only one-half of one sixty-fourth of an inch. On the other hand, if the smallest division on the ruler is one-fourth of an inch, then a measurement of $3\frac{1}{4}$ inches contains a probable error of one-eighth of an inch.

ACCURACY

Even though a number may be very precise, which indicates that it was measured with an instrument having closely spaced divisions, it may not be very accurate. The accuracy of a measurement depends upon the relative size of the probable error when compared with the quantity being measured. For example, a distance of 25 yards on a pistol range may be measured carefully enough to be correct to the nearest inch. Since there are 900 inches in 25 yards, this measurement is between 899.5 inches and 900.5 inches. When compared with the total of 900 inches, the 0.5-inch probable error is not very great.

On the other hand, a length of pipe may be

measured rather precisely and found to be 3.2 inches long. The probable error here is 0.05 inch, and this measurement is thus more precise than that of the pistol range mentioned before. To compare the accuracy of the two measurements, we note that 0.05 inch out of a total of 3.2 inches is the same as 0.5 inch out of 32 inches. Comparing this with the figure obtained in the other example (0.5 inch out of 900), we conclude that the more precise measurement is actually the less accurate of the two measurements considered.

It is important to realize that the location of the decimal point has no bearing on the accuracy of the number. For example, 1.25 dollars represents exactly the same amount of money as 125 cents. These are equally accurate ways of representing the same quantity, despite the fact that the decimal point is placed differently.

Practice problems. In each of the following problems, determine which number of each pair is more accurate and which is more precise:

1. 3.72 inches or 2,417 feet

2. 2.5 inches or 17.5 inches

3. $5\frac{3}{4}$ inches or $12\frac{7}{8}$ inches

4. 34.2 seconds or 13 seconds

Answers:

1. 3.72 inches is more precise.
   2,417 feet is more accurate.

2. The numbers are equally precise.
   17.5 inches is more accurate.

3. $12\frac{7}{8}$ inches is more precise and more accurate.

4. 34.2 seconds is more precise and more accurate.

Percent of Error

The accuracy of a measurement is determined by the RELATIVE ERROR. The relative error is the ratio between the probable error and the quantity being measured. This ratio is simply the fraction formed by using the probable error as the numerator and the measurement itself as the denominator. For example, suppose that a metal plate is found to be 5.4 inches long, correct to the nearest tenth of an inch. The maximum probable error is five hundredths of an inch (one-half of one tenth of an inch) and the relative error is found as follows:

$$\frac{\text{probable error}}{\text{measured value}} = \frac{0.05}{5.4}$$

$$= \frac{5}{540}$$

Thus the relative error is 5 parts out of 540.

Relative error is usually expressed as PERCENT OF ERROR. When the denominator of the fraction expressing the error ratio is divided into the numerator, a decimal is obtained. This decimal, converted to percent, gives the percent of error. For example, the error in the foregoing problem could be stated as 0.93

percent, since the ratio 5/540 reduces to 0.0093 (rounded off) in decimal form.

Significant Digits

The accuracy of a measurement is often described in terms of the number of significant digits used in expressing it. If the digits of a number resulting from measurement are examined one by one, beginning with the left-hand digit, the first digit that is not 0 is the first significant digit. For example, 2345 has four significant digits and 0.023 has only two significant digits.

The digits 2 and 3 in a measurement such as 0.023 inch signify how many thousandths of an inch comprise the measurement. The 0 s are of no significance in specifying the number of thousandths in the measurement; their presence is required only as "place holders" in placing the decimal point.

A rule that is often used states that the significant digits in a number begin with the first nonzero digit (counting from left to right) and end with the last digit. This implies that 0 can be a significant digit if it is not the first digit in the number. For example, 0.205 inch is a measurement having three significant digits. The 0 between the 2 and the 5 is significant because it is a part of the number specifying how many thousandths are in the measurement.

The rule stated in the foregoing paragraph fails to classify final 0 s on the right. For example, in a number such as 4,700, the number of significant digits might be two, three, or four. If the 0 s merely locate the decimal point (that is, if they show the number to be approximately forty-seven hundred rather than forty seven), then the number of significant digits is two. However, if the number 4,700 represents a number such as 4,703 rounded off to the nearest hundred, there are three significant digits. The last 0 merely locates the decimal point. If the number 4,700 represents a number such as 4,700.4 rounded off, then the number of significant digits is four.

Unless we know how a particular number was measured, it is sometimes impossible to determine whether right-hand 0 s are the result of rounding off. However, in a practical situation it is normally possible to obtain information concerning the instruments used and the degree of precision of the original data before any rounding was done.

In a number such as 49.30 inches, it is reasonable to assume that the 0 in the hundredths place would not have been recorded at all if it were not significant. In other words, the instrument used for the measurement can be read to the nearest hundredth of an inch. The 0 on the right is thus significant. This conclusion can be reached another way by observing that the 0 in 49.30 is not needed as a place holder in placing the decimal point. Therefore its presence must have some other significance.

The facts concerning significant digits may be summarized as follows:

1. Digits other than 0 are always significant.

2. Zero is significant when it falls between significant digits.

3. Any final 0 to the right of the decimal point is significant.

4. When a 0 is present only as a place holder for locating the decimal point, it is not significant.

5. The following categories comprise the significant digits of any measurement number:

a. The first nonzero left-hand digit is significant.

b. The digit which indicates the precision of the number is significant. This is the digit farthest to the right, except when the right-hand digit is 0. If it is 0, it may be only a place holder when the number is an integer.

c. All digits between significant digits are significant.

Practice problems. Determine the percent of error and the number of significant digits in each of the following measurements:

1. 5.4 feet

2. 0.00042 inch

3. 4.17 sec

4. 147.50 miles

Answers:

1. Percent of error: 0.93%
   Significant digits: 2

2. Percent of error: 1.19%
   Significant digits: 2

3. Percent of error: 0.12%
   Significant digits: 3

4. Percent of error: 0.0034%
   Significant digits: 5

CALCULATING WITH APPROXIMATE NUMBERS

The concepts of precision and accuracy form the basis for the rules which govern calculation with approximate numbers (numbers resulting from measurement).

Addition and Subtraction

A sum or difference can never be more precise than the least precise number in the calculation. Therefore, before adding or subtracting approximate numbers, they should be rounded to the same degree of precision. The more precise numbers are all rounded to the precision of the least precise number in the group to be combined. For example, the numbers 2.95, 32.7, and 1.414 would be rounded to tenths before adding as follows:

$$
\begin{array}{r}
3.0 \\
32.7 \\
\underline{1.4}
\end{array}
$$

Multiplication and Division

When two numbers are multiplied, the result often has several more digits than either of the original factors. Division also frequently produces more digits in the quotient than the original data possessed, if the division is "carried out" to several decimal places. Results such as these appear to have more significant digits than the original measurements from which they came, giving the false impression of greater accuracy than is justified. In order to correct this situation, the following rule is used:

In order to multiply or divide two approximate numbers having an equal number of significant digits, round the answer to the same number of significant digits as are shown in the original data. If one of the original factors has more significant digits than the other, round the more accurate number before multiplying. It should be rounded to one more significant digit than appears in the less accurate number; the extra digit protects the answer from the effects of multiple rounding. After performing the multiplication or division, round the result to the same number of significant digits as are shown in the less accurate of the original factors.

Practice problems:

1. Find the sum of the sides of a triangle in which the lengths of the three sides are as follows: 2.5 inches, 3.72 inches, and 4.996 inches.

2. Find the product of the length and width of a rectangle which is 2.95 feet long and 0.9046 foot wide.

Answers:

1. 11.2 inches

2. 2.67 square feet

MICROMETERS AND VERNIERS

Closely associated with the study of decimals is a measuring instrument known as a micrometer. The ordinary micrometer is capable of measuring accurately to one-thousandth of an inch. One-thousandth of an inch is about the thickness of a human hair or a thin sheet of paper. The parts of a micrometer are shown in figure 6-1.

MICROMETER SCALES

The spindle and the thimble move together. The end of the spindle (hidden from view in figure 6-1) is a screw with 40 threads per inch. Consequently, one complete turn of the thimble moves the spindle one-fortieth of an inch or

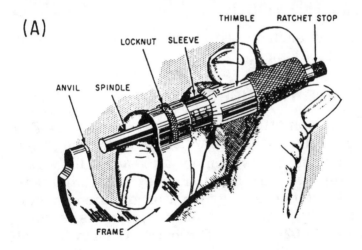

Figure 6-1.—(A) Parts of a micrometer; (B) micrometer scales.

0.025 inch since $\frac{1}{40}$ is equal to 0.025. The sleeve has 40 markings to the inch. Thus each space between the markings on the sleeve is also 0.025 inch. Since 4 such spaces are 0.1 inch (that is, 4 x 0.025), every fourth mark is labeled in tenths of an inch for convenience in reading. Thus, 4 marks equal 0.1 inch, 8 marks equal 0.2 inch, 12 marks equal 0.3 inch, etc.

To enable measurement of a partial turn, the beveled edge of the thimble is divided into 25 equal parts. Thus each marking on the thimble is $\frac{1}{25}$ of a complete turn, or $\frac{1}{25}$ of $\frac{1}{40}$ of an inch. Multiplying $\frac{1}{25}$ times 0.025 inch, we find that each marking on the thimble represents 0.001 inch.

READING THE MICROMETER

It is sometimes convenient when learning to read a micrometer to write down the component parts of the measurement as read on the scales and then to add them. For example, in figure 6-1 (B) there are two major divisions visible (0.2 inch). One minor division is showing clearly (0.025 inch). The marking on the thimble nearest the horizontal or index line of the sleeve is the second marking (0.002 inch). Adding these parts, we have

$$
\begin{array}{r}
0.200 \\
0.025 \\
0.002 \\
\hline
0.227
\end{array}
$$

Thus, the reading is 0.227 inch. As explained previously, this is read verbally as "two hundred twenty-seven thousandths." A more skillful method of reading the scales is to read all digits as thousandths directly and to do any adding mentally. Thus, we read the major division on the scale as "two hundred thousandths" and the minor division is added on mentally. The mental process for the above setting then would be "two hundred twenty-five; two hundred twenty-seven thousandths."

Practice problems:

1. Read each of the micrometer settings shown in figure 6-2.

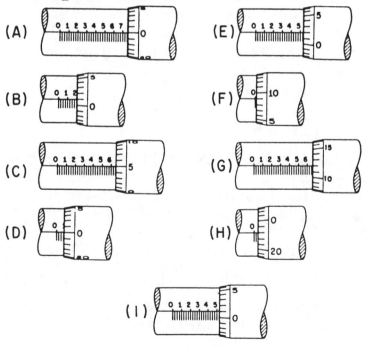

Figure 6-2.—Micrometer settings.

Answers:

1. (A) 0.750   (F) 0.009
   (B) 0.201   (G) 0.662
   (C) 0.655   (H) 0.048
   (D) 0.075   (I) 0.526
   (E) 0.527

VERNIER

Sometimes the marking on the thimble of the micrometer does not fall directly on the index line of the sleeve. To make possible readings even smaller than thousandths, an ingenious device is introduced in the form of an additional scale. This scale, called a VERNIER, was named after its inventor, Pierre Vernier. The vernier makes possible accurate readings to the ten-thousandth of an inch.

Principles of the Vernier

Suppose a ruler has markings every tenth of an inch but it is desired to read accurately to hundredths. A separate, freely sliding vernier scale (fig. 6-3) is added to the ruler. It has 10 markings on it that take up the same distance as 9 markings on the ruler scale. Thus, each space on the vernier is $\frac{1}{10}$ of $\frac{9}{10}$ inch, or $\frac{9}{100}$ inch. How much smaller is a space on the vernier than a space on the ruler? The ruler space is $\frac{1}{10}$ inch, or $\frac{10}{100}$ inch, and the vernier

182

space is $\frac{9}{100}$ inch. The vernier space is smaller by the difference between these two numbers, as follows:

$$\frac{10}{100} - \frac{9}{100} = \frac{1}{100}$$

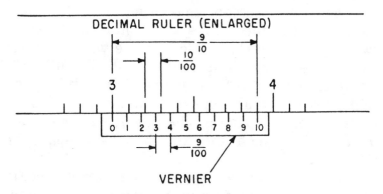

Figure 6-3.—Vernier scale.

Each vernier space is $\frac{1}{100}$ inch smaller than a ruler space.

As an example of the use of the vernier scale, suppose that we are measuring the steel bar shown in figure 6-4. The end of the bar almost reaches the 3-inch mark on the ruler, and we estimate that it is about halfway between 2.9 inches and 3.0 inches.

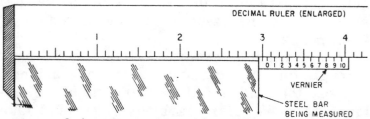

Figure 6-4.—Measuring with a vernier.

The 0 on the vernier scale is spaced the distance of exactly one ruler mark (in this case, one tenth of an inch) from the left hand end of the vernier. Therefore the 0 is at a position between ruler marks which is comparable to the position of the end of the bar. In other words, the 0 on the vernier is about halfway between two adjacent marks on the ruler, just as the end of the bar is about halfway between two adjacent marks. The 1 on the vernier scale is a little closer to alinement with an adjacent ruler mark; in fact, it is one hundredth of an inch closer to alinement than the 0. This is because each space on the vernier is one hundredth of an inch shorter than each space on the ruler.

Each successive mark on the vernier scale is one hundredth of an inch closer to alinement than the preceding mark, until finally alinement is achieved at the 5 mark. This means that the 0 on the vernier must be five hundredths of an inch from the nearest ruler mark, since five increments, each one hundredth of an inch in size, were used before a mark was found in alinement.

We conclude that the end of the bar is five hundredths of an inch from the 2.9 mark on the ruler, since its position between marks is exactly comparable to that of the 0 on the vernier scale. Thus the value of our measurement is 2.95 inches.

The foregoing example could be followed through for any distance between markings. Suppose the 0 mark fell seven tenths of the distance between ruler markings. It would take seven vernier markings, a loss of one-hundredth of an inch each time, to bring the marks in line at 7 on the vernier.

The vernier principle may be used to get fine linear readings, angular readings, etc. The principle is always the same. The vernier has one more marking than the number of markings on an equal space of the conventional scale of the measuring instrument. For example, the vernier caliper (fig. 6-5) has 25 markings on the vernier for 24 on the caliper scale. The caliper is marked off to read to fortieths (0.025) of an inch, and the vernier extends the accuracy to a thousandth of an inch.

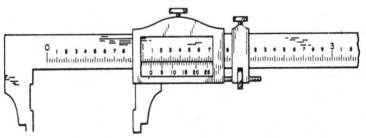

Figure 6-5.—A vernier caliper.

Vernier Micrometer

By adding a vernier to the micrometer, it is possible to read accurately to one ten-thousandth of an inch. The vernier markings are on the sleeve of the micrometer and are parallel to the thimble markings. There are 10 divisions on the vernier that occupy the same space as 9 divisions on the thimble. Since a thimble space is one thousandth of an inch, a vernier space is

185

$\dfrac{1}{10}$ of $\dfrac{9}{1000}$ inch, or $\dfrac{9}{10000}$ inch. It is $\dfrac{1}{10000}$ inch less than a thimble space. Thus, as in the preceding explanation of verniers, it is possible to read the nearest ten-thousandth of an inch by reading the vernier digit whose marking coincides with a thimble marking.

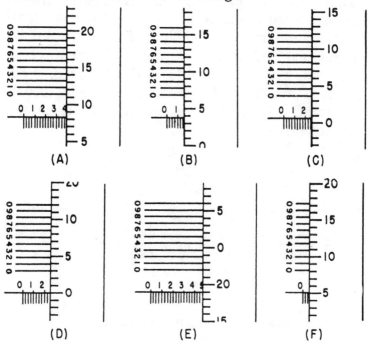

Figure 6-6.—Vernier micrometer settings.

In figure 6-6 (A), the last major division showing fully on the sleeve index is 3. The third minor division is the last mark clearly showing (0.075). The thimble division nearest and below the index is the 8 (0.008). The vernier marking that matches a thimble marking is the fourth (0.0004). Adding them all together, we have,

186

```
0.3000
0.0750
0.0080
0.0004
―――――
0.3834
```

The reading is 0.3834 inch. With practice these readings can be made directly from the micrometer, without writing the partial readings.

Practice problems:

1. Read the micrometer settings in figure 6-6.

Answers:

1. (A) See the foregoing example.

    (B) 0.1539   (E) 0.4690

    (C) 0.2507   (F) 0.0552

    (D) 0.2500

CHAPTER 7

EXPONENTS AND RADICALS

The operation of raising a number to a power is a special case of multiplication in which the factors are all equal. In examples such as $4^2 = 4 \times 4 = 16$ and $5^3 = 5 \times 5 \times 5 = 125$, the number 16 is the second power of 4 and the number 125 is the third power of 5. The expression 5^3 means that three 5 s are to be multiplied together. Similarly, 4^2 means 4×4. The first power of any number is the number itself. The power is the number of times the number itself is to be taken as a factor.

The process of finding a root is the inverse of raising a number to a power. A root is a special factor of a number, such as 4 in the expression $4^2 = 16$. When a number is taken as a factor two times, as in the expression $4 \times 4 = 16$, it is called a square root. Thus, 4 is a square root of 16. By the same reasoning, 2 is a cube root of 8, since $2 \times 2 \times 2$ is equal to 8. This relationship is usually written as $2^3 = 8$.

POWERS AND ROOTS

A power of a number is indicated by an EXPONENT, which is a number in small print placed to the right and toward the top of the number. Thus, in $4^3 = 64$, the number 3 is the EXPONENT of the number 4. The exponent 3 indicates that the number 4, called the BASE, is to be raised to its third power. The expression is read "4 to the third power (or 4 cubed) equals 64." Similarly, $5^2 = 25$ is read "5 to the second power (or 5 squared) equals 25." Higher powers are read according to the degree indicated; for example, "fourth power," "fifth power," etc.

When an exponent occurs, it must always be written unless its value is 1. The exponent 1 usually is not written, but is understood. For example, the number 5 is actually 5^1. When we work with exponents, it is important to remember that any number that has no written exponent really has an exponent equal to 1.

A root of a number can be indicated by placing a radical sign, $\sqrt{}$, over the number and showing the root by placing a small number within the notch of the radical sign. Thus, $\sqrt[3]{64}$ indicates the cube root of 64, and $\sqrt[5]{32}$ indicates the fifth root of 32. The number that indicates the root is called the INDEX of the root. In the case of the square root, the index, 2, usually is not shown. When a radical has no index, the square root is understood to be the one desired. For example, $\sqrt{36}$ indicates the square root of 36. The line above the number whose root is to be found is a symbol of grouping called the vinculum. When the radical symbol is used, a vinculum, long enough to extend over the entire expression whose root is to be found, should be attached.

Practice problems. Raise to the indicated power or find the root indicated.

1. 2^3 2. 6^2 3. 4^3 4. 25^3

5. $\sqrt{16}$ 6. $\sqrt[3]{8}$ 7. $\sqrt[3]{125}$ 8. $\sqrt[5]{32}$

Answers:

1. 8   2. 36   3. 64   4. 15,625
5. 4   6. 2   7. 5   8. 2

NEGATIVE INTEGERS

Raising to a power is multiplication in which all the numbers being multiplied together are equal. The sign of the product is determined, as in ordinary multiplication, by the number of minus signs. The number of minus signs is odd or even, depending on whether the exponent of the power is odd or even. For example, in the problem

$$(-2)^3 = (-2)(-2)(-2) = -8$$

there are three minus signs. The result is negative. In

$$(-2)^6 = 64$$

there are six minus signs. The result is positive.

Thus, when the exponent of a negative number is odd, the power is negative; when the exponent is even, the power is positive.

190

As other examples, consider the following:

$$(-3)^4 = 81$$

$$\left(-\frac{2}{5}\right)^3 = -\frac{8}{125}$$

$$(-2)^8 = 256$$

$$(-1)^5 = -1$$

Positive and negative numbers belong to the class called REAL NUMBERS. The square of a real number is positive. For example, $(-7)^2 = 49$ and $7^2 = 49$. The expression $(-7)^2$ is read "minus seven squared." Note that either seven squared or minus seven squared gives us +49. We cannot obtain -49 or any other negative number by squaring any real number, positive or negative.

Since there is no real number whose square is a negative number, it is sometimes said that the square root of a negative number does not exist. However, an expression under a square root sign may take on negative values. While the square root of a negative number cannot actually be found, it can be indicated.

The indicated square root of a negative number is called an IMAGINARY NUMBER. The number $\sqrt{-7}$, for example, is said to be imaginary. It is read "square root of minus seven." Imaginary numbers are discussed in chapter 15 of this course.

FRACTIONS

We recall that the exponent of a number tells

the number of times that the number is to be taken as a factor. A fraction is raised to a power by raising the numerator and the denominator separately to the power indicated. The expression $\left(\frac{3}{7}\right)^2$ means $\frac{3}{7}$ is used twice as a factor. Thus,

$$\left(\frac{3}{7}\right)^2 = \frac{3}{7} \times \frac{3}{7} = \frac{3^2}{7^2}$$

$$= \frac{9}{49}$$

Similarly,

$$\left(-\frac{1}{5}\right)^2 = \frac{1}{25}$$

Since a minus sign can occupy any one of three locations in a fraction, notice that evaluating $\left(-\frac{1}{5}\right)^2$ is equivalent to

$$(-1)^2 \left(\frac{1}{5}\right)^2 \text{ or } \frac{(-1)^2}{5^2} \text{ or } \frac{1^2}{(-5)^2}$$

The process of taking a root of a number is the inverse of the process of raising the number to a power, and the method of taking the root of a fraction is similar. We may simply take the root of each term separately and write the result as a fraction. Consider the following examples:

$$1. \sqrt{\frac{36}{49}} = \frac{\sqrt{36}}{\sqrt{49}} = \frac{6}{7}$$

2. $\sqrt[3]{\dfrac{8}{125}} = \dfrac{\sqrt[3]{8}}{\sqrt[3]{125}} = \dfrac{2}{5}$

Practice problems. Find the values for the indicated operations:

1. $\left(\dfrac{1}{3}\right)^2$ 2. $\left(\dfrac{3}{4}\right)^2$ 3. $\left(\dfrac{6}{5}\right)^2$ 4. $\left(\dfrac{2}{3}\right)^3$

5. $\sqrt{\dfrac{16}{36}}$ 6. $\sqrt{\dfrac{16}{25}}$ 7. $\sqrt[3]{\dfrac{8}{27}}$ 8. $\sqrt{\dfrac{9}{49}}$

Answers:

1. 1/9   2. 9/16   3. 36/25   4. 8/27

5. 4/6   6. 4/5   7. 2/3   8. 3/7

DECIMALS

When a decimal is raised to a power, the number of decimal places in the result is equal to the number of places in the decimal multiplied by the exponent. For example, consider $(0.12)^3$. There are two decimal places in 0.12 and 3 is the exponent. Therefore, the number of places in the power will be $3(2) = 6$. The result is as follows:

$$(0.12)^3 = 0.001728$$

The truth of this rule is evident when we recall the rule for multiplying decimals. Part of the rule states: Mark off as many decimal places in the product as there are decimal places in the factors together. If we carry out the multiplication, (0.12) x (0.12) x (0.12), it is obvious that there are six decimal places in the three factors together. The rule can be shown for any decimal raised to any power by simply carrying out the multiplication indicated by the exponent.

Consider these examples:

$$(1.4)^2 = 1.96$$
$$(0.12)^2 = 0.0144$$
$$(0.4)^3 = 0.064$$
$$(0.02)^2 = 0.0004$$
$$(0.2)^2 = 0.04$$

Finding a root of a number is the inverse of raising a number to a power. To determine the number of decimal places in the root of a perfect power, we divide the number of decimal places in the radicand by the index of the root. Notice that this is just the opposite of what was done in raising a number to a power.

Consider $\sqrt{0.0625}$. The square root of 625 is 25. There are four decimal places in the radicand, 0.0625, and the index of the root is 2. Therefore, $4 \div 2 = 2$ is the number of decimal places in the root. We have

$$\sqrt{0.0625} = 0.25$$

Similarly,
$$\sqrt{1.69} = 1.3$$
$$\sqrt[3]{0.027} = 0.3$$
$$\sqrt[3]{1.728} = 1.2$$
$$\sqrt[4]{0.0001} = 0.1$$

194

LAWS OF EXPONENTS

All of the laws of exponents may be developed directly from the definition of exponents. Separate laws are stated for the following five cases:
1. Multiplication.
2. Division.
3. Power of a power.
4. Power of a product.
5. Power of quotient.

MULTIPLICATION

To illustrate the law of multiplication, we examine the following problem:

$$4^3 \times 4^2 = ?$$

Recalling that 4^3 means $4 \times 4 \times 4$ and 4^2 means 4×4, we see that 4 is used as a factor five times. Therefore $4^3 \times 4^2$ is the same as 4^5. This result could be written as follows:

$$4^3 \times 4^2 = 4 \times 4 \times 4 \times 4 \times 4$$
$$= 4^5$$

Notice that three of the five 4's came from the expression 4^3, and the other two 4's came from the expression 4^2. Thus we may rewrite the problem as follows:

$$4^3 \times 4^2 = 4^{(3+2)}$$
$$= 4^5$$

The law of exponents for multiplication may be stated as follows: To multiply two or more powers having the same base, add the exponents and raise the common base to the sum of the exponents. This law is further illustrated by the following examples:

$$2^3 \times 2^4 = 2^7$$
$$3 \times 3^2 = 3^3$$
$$15^4 \times 15^2 = 15^6$$
$$10^2 \times 10^{0.5} = 10^{2.5}$$

COMMON ERRORS

It is important to realize that the base must be the same for each factor, in order to apply the laws of exponents. For example, $2^3 \times 3^2$ is neither 2^5 nor 3^5. There is no way to apply the law of exponents to a problem of this kind. Another common mistake is to multiply the bases together. For example, this kind of error in the foregoing problem would imply that $2^3 \times 3^2$ is equivalent to 6^5, or 7776. The error of this may be proved as follows:

$$2^3 \times 3^2 = 8 \times 9$$
$$= 72$$

DIVISION

The law of exponents for division may be developed from the following example:

$$6^7 \div 6^5 = \frac{\cancel{6} \times \cancel{6} \times \cancel{6} \times \cancel{6} \times \cancel{6} \times 6 \times 6}{\cancel{6} \times \cancel{6} \times \cancel{6} \times \cancel{6} \times \cancel{6}}$$
$$= 6^2$$

Cancellation of the five 6 s in the divisor with five of the 6 s in the dividend leaves only two 6 s, the product of which is 6^2.

This result can be reached directly by noting that 6^2 is equivalent to $6^{(7-5)}$. In other words, we have the following:

$$6^7 \div 6^5 = 6^{(7-5)}$$
$$= 6^2$$

Therefore the law of exponents for division is as follows: To divide one power into another having the same base, subtract the exponent of the divisor from the exponent of the dividend. Use the number resulting from this subtraction as the exponent of the base in the quotient.

Use of this rule sometimes produces a negative exponent or an exponent whose value is 0. These two special types of exponents are discussed later in this chapter.

POWER OF A POWER

Consider the example $(3^2)^4$. Remembering that an exponent shows the number of times the base is to be taken as a factor and noting in this case that 3^2 is considered the base, we have

$$(3^2)^4 = 3^2 \cdot 3^2 \cdot 3^2 \cdot 3^2$$

Also in multiplication we add exponents. Thus,

$$3^2 \cdot 3^2 \cdot 3^2 \cdot 3^2 = 3^{(2+2+2+2)} = 3^8$$

Therefore,

$$(3^2)^4 = 3^{(4 \times 2)}$$
$$= 3^8$$

The laws of exponents for the power of a power may be stated as follows: To find the power of a power, multiply the exponents. It should be noted that this case is the only one in which multiplication of exponents is performed.

POWER OF A PRODUCT

Consider the example $(3 \cdot 2 \cdot 5)^3$. We know that

$$(3 \cdot 2 \cdot 5)^3 = (3 \cdot 2 \cdot 5)(3 \cdot 2 \cdot 5)(3 \cdot 2 \cdot 5)$$

Thus 3, 2, and 5 appear three times each as factors, and we can show this with exponents as 3^3, 2^3, and 5^3. Therefore,

$$(3 \cdot 2 \cdot 5)^3 = 3^3 \cdot 2^3 \cdot 5^3$$

The law of exponents for the power of a product is as follows: The power of a product is equal to the product obtained when each of the original factors is raised to the indicated power and the resulting powers are multiplied together.

POWER OF A QUOTIENT

The law of exponents for a power of an indicated quotient may be developed from the following example:

$$\left(\frac{2}{3}\right)^3 = \frac{2}{3} \cdot \frac{2}{3} \cdot \frac{2}{3}$$

$$= \frac{2 \cdot 2 \cdot 2}{3 \cdot 3 \cdot 3}$$

$$= \frac{2^3}{3^3}$$

Therefore,

$$\left(\frac{2}{3}\right)^3 = \frac{2^3}{3^3}$$

The law is stated as follows: The power of a quotient is equal to the quotient obtained when the dividend and divisor are each raised to the indicated power separately, before the division is performed.

Practice problems. Raise each of the following expressions to the indicated power:

1. $\left(3^2 \cdot 2^3\right)^2$ 3. $\left(\frac{3 \cdot 2}{5 \cdot 6}\right)^3$ 5. $\frac{5^3}{5}$

2. $3^5 \div 3^2$ 4. $(-3^2)^3$ 6. $(3 \cdot 2 \cdot 7)^2$

Answers:

1. $3^4 \times 2^6 = 5,184$

2. 27

3. $\frac{1}{125}$

4. $\left[(-3)^2\right]^3 = 729$

5. 25

6. $9 \cdot 4 \cdot 49 = 1{,}764$

SPECIAL EXPONENTS

Thus far in this discussion of exponents, the emphasis has been on exponents which are positive integers. There are two types of exponents which are not positive integers, and two which are treated as special cases even though they may be considered as positive integers.

ZERO AS AN EXPONENT

Zero occurs as an exponent in the answer to a problem such as $4^3 \div 4^3$. The law of exponents for division states that the exponents are to be subtracted. This is illustrated as follows:

$$\frac{4^3}{4^3} = 4^{(3-3)} = 4^0$$

Another way of expressing the result of dividing 4^3 by 4^3 is to use the fundamental axiom which states that any number divided by itself is 1. In order for the laws of exponents to hold true in all cases, this must also be true when any number raised to a power is divided by itself. Thus, $4^3/4^3$ must equal 1.

Since $4^3/4^3$ has been shown to be equal to both 4^0 and 1, we are forced to the conclusion that $4^0 = 1$.

By the same reasoning,

$$\frac{5}{5} = 5^{1-1} = 5^0$$

Also,

$$\frac{5}{5} = 1$$

Therefore,

$$5^0 = 1$$

Thus we see that any number divided by itself results in a 0 exponent and has a value of 1. By definition then, any number (other than zero) raised to the zero power equals 1. This is further illustrated in the following examples:

$$3^0 = 1$$
$$400^0 = 1$$
$$0.02^0 = 1$$
$$\left(\frac{1}{5}\right)^0 = 1$$
$$(\sqrt{3})^0 = 1$$

ONE AS AN EXPONENT

The number 1 arises as an exponent sometimes as a result of division. In the example $\frac{5^3}{5^2}$ we subtract the exponents to get

$$5^{3-2} = 5^1$$

This problem may be worked another way as follows:

$$\frac{5^3}{5^2} = \frac{\cancel{5} \cdot \cancel{5} \cdot 5}{\cancel{5} \cdot \cancel{5}} = 5$$

Therefore,

$$5^1 = 5$$

We conclude that any number raised to the first power is the number itself. The exponent 1 usually is not written but is understood to exist.

NEGATIVE EXPONENTS

If the law of exponents for division is extended to include cases where the exponent of the denominator is larger, negative exponents arise. Thus,

$$\frac{3^2}{3^5} = 3^{2-5} = 3^{-3}$$

Another way of expressing this problem is as follows:

$$\frac{3^2}{3^5} = \frac{\cancel{3} \cdot \cancel{3}}{\cancel{3} \cdot \cancel{3} \cdot 3 \cdot 3 \cdot 3} = \frac{1}{3^3}$$

Therefore,

$$3^{-3} = \frac{1}{3^3}$$

We conclude that a number N with a negative exponent is equivalent to a fraction having the following form: Its numerator is 1; its denominator is N with a positive exponent whose absolute value is the same as the absolute value of the original exponent. In symbols, this rule may be stated as follows:

$$N^{-a} = \frac{1}{N^a}$$

Also,

$$\frac{1}{N^{-a}} = N^a$$

The following examples further illustrate the rule:

$$5^{-1} = \frac{1}{5}$$

$$6^{-2} = \frac{1}{6^2}$$

$$4^{-12} = \frac{1}{4^{12}}$$

$$\frac{1}{3^{-2}} = 3^2$$

Notice that the sign of an exponent may be changed by merely moving the expression which contains the exponent to the other position in the fraction. The sign of the exponent is changed as this move is made. For example,

$$\frac{1}{10^{-2}} = 1 \div \frac{1}{10^2}$$

$$= 1 \times \frac{10^2}{1}$$

Therefore,

$$\frac{1}{10^{-2}} = \frac{10^2}{1}$$

By using the foregoing relationship, a problem such as $3 \div 5^{-4}$ may be simplified as follows:

$$\frac{3}{5^{-4}} = 3 \times \frac{1}{5^{-4}}$$

$$= 3 \times \frac{5^4}{1}$$

$$= 3 \times 5^4$$

FRACTIONAL EXPONENTS

Fractional exponents obey the same laws as do integral exponents. For example,

$$4^{1/2} \times 4^{1/2} = 4^{(1/2 + 1/2)}$$

$$= 4^{2/2}$$

$$= 4^1 = 4$$

Another way of expressing this would be

$$4^{1/2} \times 4^{1/2} = (4^{1/2})^2$$

$$= 4^{(1/2 \times 2)}$$

$$= 4^1 = 4$$

Observe that the number $4^{1/2}$, when squared in the foregoing example, produced the number 4 as an answer. Recalling that a square root of a number N is a number x such that $x^2 = N$, we conclude that $4^{1/2}$ is equivalent to $\sqrt{4}$. Thus we have a definition, as follows: A fractional exponent of the form $1/r$ indicates a root, the index of which is r. This is further illustrated in the following examples:

$$2^{1/2} = \sqrt{2}$$

$$4^{1/3} = \sqrt[3]{4}$$

$$6^{2/3} = (6^{1/3})^2 = (\sqrt[3]{6})^2$$

Also,

$$6^{2/3} = (6^2)^{1/3} = \sqrt[3]{36}$$

Notice that in an expression such as $8^{2/3}$ we can either find the cube root of 8 first or square 8 first, as shown by the following example:

$$(8^{1/3})^2 = 2^2 = 4 \text{ and } (8^2)^{1/3} = \sqrt[3]{64} = 4$$

All the numbers in the evaluation of $8^{2/3}$ remain small if the cube root is found before raising the number to the second power. This order of operation is particularly desirable in evaluating a number like $64^{5/6}$. If 64 were first raised to the fifth power, a large number would result. It would require a great deal of unnecessary effort to find the sixth root of 64^5. The result is obtained easily, if we write

$$64^{5/6} = (64^{1/6})^5 = 2^5 = 32$$

If an improper fraction occurs in an exponent, such as $7/3$ in the expression $2^{7/3}$, it is customary to keep the fraction in that form rather than express it as a mixed number. In fraction form an exponent shows immediately what power is intended and what root is intended. However, $2^{7/3}$ can be expressed in another form and simplified by changing the improper fraction to a mixed number and writing the fractional part in the radical form as follows:

$$2^{7/3} = 2^{2 + 1/3} = 2^2 \cdot 2^{1/3} = 4 \sqrt[3]{2}$$

The law of exponents for multiplication may be combined with the rule for fractional exponents to solve problems of the following type:

PROBLEM: Evaluate the expression $4^{2.5}$.

SOLUTION:
$$4^{2.5} = 4^2 \times 4^{0.5}$$
$$= 16 \times 4^{1/2}$$
$$= 16 \times 2$$
$$= 32$$

Practice problems:

1. Perform the indicated division: $\dfrac{2}{2^{1/3}}$

2. Find the product: $7^{2/5} \times 7^{1/10} \times 7^{3/10}$

3. Rewrite with a positive exponent and simplify: $9^{-1/2}$

4. Evaluate $100^{3/2}$

5. Evaluate $(8^0)^5$

Answers:

1. $2^{3/3} \div 2^{1/3} = \sqrt[3]{4}$
2. $7^{8/10}$

3. $\dfrac{1}{9^{1/2}} = \dfrac{1}{3}$

4. 1,000

5. 1

SCIENTIFIC NOTATION AND POWERS OF 10

Technicians, engineers, and others engaged in scientific work are often required to solve problems involving very large and very small numbers. Problems such as

$$\frac{22,684 \times 0.00189}{0.0713 \times 83 \times 7}$$

are not uncommon. Solving such problems by the rules of ordinary arithmetic is laborious and time consuming. Moreover, the tedious arithmetic process lends itself to operational errors. Also there is difficulty in locating the decimal point in the result. These difficulties can be greatly reduced by a knowledge of the powers of 10 and their use.

The laws of exponents form the basis for calculation using powers of 10. The following list includes several decimals and whole numbers expressed as powers of 10:

10,000	$= 10^4$	0.1	$= 10^{-1}$
1,000	$= 10^3$	0.01	$= 10^{-2}$
100	$= 10^2$	0.001	$= 10^{-3}$
10	$= 10^1$	0.0001	$= 10^{-4}$
1	$= 10^0$		

The concept of scientific notation may be demonstrated as follows:

$$60,000 = 6.0000 \times 10,000$$
$$= 6 \times 10^4$$
$$538 = 5.38 \times 100$$
$$= 5.38 \times 10^2$$

Notice that the final expression in each of the foregoing examples involves a number between 1 and 10, multiplied by a power of 10. Furthermore, in each case the exponent of the power of 10 is a number equal to the number of digits between the new position of the decimal point and the original position (understood) of the decimal point.

We apply this reasoning to write any number in scientific notation; that is, as a number between 1 and 10 multiplied by the appropriate power of 10. The appropriate power of 10 is found by the following mechanical steps:

1. Shift the decimal point to standard position, which is the position immediately to the right of the first nonzero digit.

2. Count the number of digits between the new position of the decimal point and its original position. This number indicates the value of the exponent for the power of 10.

3. If the decimal point is shifted to the left, the sign of the exponent of 10 is positive; if the decimal point is shifted to the right, the sign of the exponent is negative.

The validity of this rule, for those cases in which the exponent of 10 is negative, is demonstrated as follows:

$$0.00657 = 6.57 \times 0.001$$
$$= 6.57 \times 10^{-3}$$
$$0.348 = 3.48 \times 0.1$$
$$= 3.48 \times 10^{-1}$$

Further examples of the use of scientific notation are given as follows:

$$543,000,000 = 5.43 \times 10^{8}$$
$$186 = 1.86 \times 10^{2}$$
$$243.01 = 2.4301 \times 10^{2}$$
$$0.0000007 = 7 \times 10^{-7}$$
$$0.00023 = 2.3 \times 10^{-4}$$

Multiplication Using Powers of 10

From the law of exponents for multiplication we recall that to multiply two or more powers to the same base we add their exponents. Thus,

$$10^{4} \times 10^{2} = 10^{6}$$

We see that multiplying powers of 10 together is an application of the general rule. This is demonstrated in the following examples.

1. $$10,000 \times 100 = 10^{4} \times 10^{2}$$
$$= 10^{4 + 2}$$
$$= 10^{6}$$

2. $$0.0000001 \times 0.001 = 10^{-7} \times 10^{-3}$$
$$= 10^{-7 + (-3)}$$
$$= 10^{-10}$$

3. $10,000 \times 0.001 = 10^4 \times 10^{-3}$

$\qquad = 10^{4-3}$

$\qquad = 10$

4. $23,000 \times 500 = ?$

$\qquad 23,000 = 2.3 \times 10^4$

$\qquad 500 = 5 \times 10^2$

Therefore,

$\qquad 23,000 \times 500 = 2.3 \times 10^4 \times 5 \times 10^2$

$\qquad = 2.3 \times 5 \times 10^4 \times 10^2$

$\qquad = 11.5 \times 10^6$

$\qquad = 11,500,000$

5. $62,000 \times 0.003 \times 4,600 = ?$

$\qquad 62,000 = 6.2 \times 10^4$

$\qquad 0.0003 = 3 \times 10^{-4}$

$\qquad 4,600 = 4.6 \times 10^3$

Therefore,

$\qquad 62,000 \times 0.0003 \times 4,600 = 6.2 \times 3 \times 4.6 \times 10^4 \times 10^{-4} \times 10^3$

$\qquad = 85.56 \times 10^3$

$\qquad = 85,560$

Practice problems. Multiply, using powers of 10. For the purposes of this exercise, treat all numbers as exact numbers:

1. $10,000 \times 0.001 \times 100$

2. 0.000350 x 5,000,000 x 0.0004

3. 3,875 x 0.000032 x 3,000,000

4. 7,000 x 0.015 x 1.78

Answers:

1. 1.0×10^3

2. 7.0×10^{-1}

3. 3.72×10^5

4. 1.869×10^2

Division Using Powers of 10

The rule of exponents for division states that, for powers of the same base, the exponent of the denominator is subtracted from the exponent of the numerator. Thus,

$$\frac{10^7}{10^3} = 10^{7-3}$$

$$= 10^4$$

It should be remembered that powers may be transferred from numerator to denominator or from denominator to numerator by simply changing the sign of the exponent. The following examples illustrate the use of this rule for powers of 10:

1. $$\frac{72,000}{0.0012} = \frac{7.2 \times 10^4}{1.2 \times 10^{-3}}$$

$$= \frac{7.2}{1.2} \times 10^4 \times 10^3$$

$$= 6 \times 10^7$$

211

2. $$\frac{44 \text{ x } 10^{-4}}{11 \text{ x } 10^{-5}} = \frac{44}{11} \text{ x } 10^{-4} \text{ x } 10^{5}$$

$$= 4 \text{ x } 10$$

Combined Multiplication and Division

Using the rules already shown, multiplication and division involving powers of 10 may be combined. The usual method of solving such problems is to multiply and divide alternately until the problem is completed. For example,

$$\frac{36,000 \text{ x } 1.1 \text{ x } 0.06}{0.012 \text{ x } 2,200}$$

Rewriting this problem in scientific notation, we have

$$\frac{3.6 \text{ x } 10^{4} \text{ x } 1.1 \text{ x } 6 \text{ x } 10^{-2}}{1.2 \text{ x } 10^{-2} \text{ x } 2.2 \text{ x } 10^{3}} = \frac{3.6 \text{ x } 1.1 \text{ x } 6}{1.2 \text{ x } 2.2} \text{ x } 10$$

$$= 9 \text{ x } 10$$

$$= 90$$

Notice that the elimination of 0 s, wherever possible, simplifies the computation and makes it an easy matter to place the decimal point.

SIGNIFICANT DIGITS.—One of the most important advantages of scientific notation is the fact that it simplifies the task of determining the number of significant digits in a number. For example, the fact that the number 0.00045 has two significant digits is sometimes obscured by the presence of the 0 s. The confusion can be avoided by writing the number in scientific notation, as follows:

$$0.00045 = 4.5 \times 10^{-4}$$

Practice problems. Express the numbers in the following problems in scientific notation and round off before performing the calculation. In each problem, round off calculation numbers to one more digit than the number of significant digits in the least accurate number; round the answer to the number of significant digits in the least accurate number:

1. $\dfrac{0.000063 \times 50.4 \times 0.007213}{780 \times 0.682 \times 0.018}$

2. $\dfrac{0.015 \times 216 \times 1.78}{72 \times 0.0624 \times 0.0353}$

3. $\dfrac{0.000079 \times 0.00036}{29 \times 10^{-8}}$

Answers:

1. 2.38×10^{-6}

2. 3.64×10

3. 9.8×10^{-2}

Other Applications

The applications of powers of 10 may be broadened to include problems involving reciprocals and powers of products.

RECIPROCALS.—The following example illustrates the use of powers of 10 in the formation of a reciprocal:

$$\frac{1}{250,000 \times 300 \times 0.02}$$

$$= \frac{1}{2.5 \times 10^5 \times 3 \times 10^2 \times 2 \times 10^{-2}}$$

$$= \frac{10^{-5}}{2.5 \times 3 \times 2}$$

$$= \frac{10^{-5}}{15}$$

Rather than write the numerator as 0.00001, write it as the product of two factors, one of which may be easily divided, as follows:

$$\frac{10^{-5}}{15} = \frac{10^2 \times 10^{-7}}{15}$$

$$= \frac{100}{15} \times 10^{-7}$$

$$= 6.67 \times 10^{-7}$$

$$= 0.000000667$$

POWER OF A PRODUCT.—The following example illustrates the use of powers of 10 in finding the power of a product:

$$(80,000 \times 2 \times 10^5)^2 = (8 \times 10^4 \times 2 \times 10^5)^2$$

$$= 8^2 \times 2^2 \times (10^{4+5})^2$$

$$= 64 \times 4 \times 10^{18}$$

$$= 256 \times 10^{18}$$

$$= 2.56 \times 10^{20}$$

RADICALS

An expression such as $\sqrt{2}$, $\sqrt[3]{5}$, or $\sqrt{a + b}$ that exhibits a radical sign, is referred to as a RADICAL. We have already worked with radicals in the form of fractional exponents, but it is also frequently necessary to work with them in the radical form. The word "radical" is derived from the Latin word "radix,"which means "root." The word "radix" itself is more often used in modern mathematics to refer to the base of a number system, such as the base 2 in the binary system. However, the word "radical" is retained with its original meaning of "root."

The radical symbol ($\sqrt{\ }$) appears to be a distortion of the initial letter "r" from the word "radix." With long usage, the r gradually lost its significance as a letter and became distorted into the symbol as we use it. The vinculum helps to specify exactly which of the letters and numbers following the radical sign actually belong to the radical expression.

The number under a radical sign is the RADICAND. The index of the root (except in the case of a square root) appears in the trough of the radical sign. The index tells what root of the radicand is intended. For example, in $\sqrt[5]{32}$, the radicand is 32 and the index of the root is 5. The fifth root of 32 is intended. In $\sqrt{50}$, the square root of 50 is intended. When the index is 2, it is not written, but is understood.

If we can find one square root of a number we can always find two of them. Remember $(3)^2$ is 9 and $(-3)^2$ is also 9. Likewise $(4)^2$ and $(-4)^2$ both equal 16 and $(5)^2$ and $(-5)^2$ both equal 25. Conversely, $\sqrt{9}$ is +3 or -3, $\sqrt{16}$ is +4 or

-4, and $\sqrt{25}$ is +5 or -5. When we wish to show a number that may be either positive or negative, we may use the symbol ± which is read "plus or minus." Thus ± 3 means "plus or minus 3." Usually when a number is placed under the radical sign, only its positive root is desired and, unless otherwise specified, it is the only root that need be found.

COMBINING RADICALS

A number written in front of another number and intended as a multiplier is called a COEFFICIENT. The expression 5x means 5 times x; ay means a times y; and $7 \sqrt{2}$ means 7 times $\sqrt{2}$. In these examples, 5 is the coefficient of x, a is the coefficient of y, and 7 is the coefficient of $\sqrt{2}$.

Radicals having the same index and the same radicand are SIMILAR. Similar radicals may have different coefficients in front of the radical sign. For example, $3 \sqrt{2}$, $\sqrt{2}$, and $\frac{1}{5} \sqrt{2}$ are similar radicals. When a coefficient is not written, it is understood to be 1. Thus, the coefficient of $\sqrt{2}$ is 1. The rule for adding radicals is the same as that stated for adding denominate numbers: Add only units of the same kind. For example, we could add $2 \sqrt{3}$ and $4 \sqrt{3}$ because the "unit" in each of these numbers is the same $(\sqrt{3})$. By the same reasoning, we could not add $2 \sqrt{3}$ and $4 \sqrt{5}$ because these are not similar radicals.

Addition and Subtraction

When addition or subtraction of similar radicals is indicated, the radicals are combined by adding or subtracting their coefficients and placing the result in front of the radical. Adding $3\sqrt{2}$ and $5\sqrt{2}$ is similar to adding 3 bolts and 5 bolts. The following examples illustrate the addition and subtraction of similar radical expressions:

1. $3\sqrt{2} + 5\sqrt{2} = 8\sqrt{2}$

2. $1/2\left(\sqrt[4]{3}\right) + 1/3\left(\sqrt[4]{3}\right) = 5/6\left(\sqrt[4]{3}\right)$

3. $\sqrt{5} - 6\sqrt{5} + 2\sqrt{5} = -3\sqrt{5}$

4. $-5\sqrt[3]{7} - 2\sqrt[3]{7} + 7\sqrt[3]{7} = 0$

Example 4 illustrates a case that is sometimes troublesome. The sum of the coefficients, -5, -2, and 7, is 0. Therefore, the coefficient of the answer would be 0, as follows:

$$0(\sqrt[3]{7}) = 0 \times \sqrt[3]{7}$$

Thus the final answer is 0, since 0 multiplied by any quantity is still 0.

Practice problems. Perform the indicated operations:

1. $4\sqrt{3} - \sqrt{3} + 5\sqrt{3}$

2. $\frac{1}{2}\sqrt{6} + \sqrt{6}$

3. $\sqrt[3]{5} - 6\sqrt[3]{5}$

4. $-2\sqrt{10} - 7\sqrt{10}$

Answers

1. $8\sqrt{3}$

2. $\dfrac{3}{2}\sqrt{6}$

3. $-5\sqrt[3]{5}$

4. $-9\sqrt{10}$

Multiplication and Division

If a radical is written immediately after another radical, multiplication is intended. Sometimes a dot is placed between the radicals, but not always. Thus, either $\sqrt{7} \cdot \sqrt{11}$ or $\sqrt{7}\,\sqrt{11}$ means multiplication.

When multiplication or division of radicals is indicated, several radicals having the same index can be combined into one radical, if desired. Radicals having the same index are said to be of the SAME ORDER. For example, $\sqrt{2}$ is a radical of the second order. The radicals $\sqrt{2}$ and $\sqrt{5}$ are of the same order.

If radicals are of the same order, the radicands can be multiplied or divided and placed under one radical symbol. For example, $\sqrt{5}$ multiplied by $\sqrt{3}$ is the same as $\sqrt{5 \times 3}$. Also, $\sqrt{6}$ divided by $\sqrt{3}$ is the same as $\sqrt{6 \div 3}$. If coefficients appear before the radicals, they also must be included in the multiplication or division. This is illustrated in the following examples:

1.
$$2\sqrt{2} \cdot 3\sqrt{5} = 2 \cdot \sqrt{2} \cdot 3 \cdot \sqrt{5}$$
$$= 2 \cdot 3 \sqrt{2} \cdot \sqrt{5}$$
$$= 2 \cdot 3 \sqrt{2 \cdot 5}$$
$$= 6\sqrt{10}$$

218

2.
$$\frac{15 \sqrt{6}}{3 \sqrt{3}} = \frac{15}{3} \times \sqrt{\frac{6}{3}}$$

$$= 5 \times \sqrt{2}$$

$$= 5 \sqrt{2}$$

It is important to note that what we have said about multiplication and division does not apply to addition. A typical error is to treat the expression $\sqrt{9 + 4}$ as if it were equivalent to $\sqrt{9} + \sqrt{4}$. These expressions cannot be equivalent, since $3 + 2$ is not equivalent to $\sqrt{13}$.

FACTORING RADICALS.—A radical can be split into two or more radicals of the same order if the radicand can be factored. This is illustrated in the following examples:

1. $\sqrt{20} = \sqrt{4} \cdot \sqrt{5} = 2 \sqrt{5}$

2. $\sqrt[3]{54} = \sqrt[3]{27 \cdot 2}$

$$= \sqrt[3]{27} \cdot \sqrt[3]{2} = 3 \sqrt[3]{2}$$

3. $\dfrac{\sqrt{20}}{\sqrt{5}} = \dfrac{\sqrt{4} \cdot \sqrt{5}}{\sqrt{5}}$

$$= \sqrt{4} = 2$$

SIMPLIFYING RADICALS

Some radicals may be changed to an equivalent form that is easier to use. A radical is in its simplest form when no factor can be removed from the radical, when there is no fraction under the radical sign, and when the index of the root cannot be reduced. A factor can be removed from the radical if it occurs a number of times equal to the index of the root. The following examples illustrate this:

1. $\sqrt{28} = \sqrt{2^2 \cdot 7} = 2\sqrt{7}$
2. $\sqrt[3]{54} = \sqrt[3]{3^3 \cdot 2} = 3(\sqrt[3]{2})$
3. $\sqrt[5]{160} = \sqrt[5]{2^5 \cdot 5} = 2(\sqrt[5]{5})$

Removing a factor that occurs a number of times equal to the index of the root is equivalent to separating a radical into two radicals so that one radicand is a perfect power. The radical sign can be removed from the number that is a perfect square, cube, fourth power, etc. The root taken becomes the coefficient of the remaining radical.

In order to simplify radicals easily, it is convenient to know the squares of whole numbers up to about 25 and a few of the smaller powers of the numbers 2, 3, 4, 5, and 6. Table 7-1 shows some frequently used powers of numbers.

Referring to table 7-1 (A), we see that the series of numbers

1, 4, 9, 16, 25, 36, 49, 64, 81, 100

comprises all the perfect squares from 1 to 100 inclusive. If any one of these numbers appears under a square root symbol, the radical sign can be removed immediately. This is illustrated as follows:

$$\sqrt{25} = 5$$

$$\sqrt{81} = 9$$

A radicand such as 75, which has a perfect

square (25) as a factor, can be simplified as follows:

$$\sqrt{75} = \sqrt{25 \cdot 3}$$
$$= \sqrt{25} \cdot \sqrt{3}$$
$$= 5\sqrt{3}$$

Table 7-1.—Powers of numbers.

$1^2 = 1$	$14^2 = 196$
$2^2 = 4$	$15^2 = 225$
$3^2 = 9$	$16^2 = 256$
$4^2 = 16$	$17^2 = 289$
$5^2 = 25$	$18^2 = 324$
$6^2 = 36$	$19^2 = 361$
$7^2 = 49$	$20^2 = 400$
$8^2 = 64$	$21^2 = 441$
$9^2 = 81$	$22^2 = 484$
$10^2 = 100$	$23^2 = 529$
$11^2 = 121$	$24^2 = 576$
$12^2 = 144$	$25^2 = 625$
$13^2 = 169$	

(A)

Table 7-1.—Powers of numbers—Continued.

$2^1 = 2$	$2^3 = 8$	$2^5 = 32$	$2^7 = 128$
$2^2 = 4$	$2^4 = 16$	$2^6 = 64$	$2^8 = 256$

(B)

221

Table 7-1.—Powers of numbers—Continued.

$$3^1 = 3$$
$$3^2 = 9$$
$$3^3 = 27$$
$$3^4 = 81$$
$$3^5 = 243$$

(C)

$$4^1 = 4$$
$$4^2 = 16$$
$$4^3 = 64$$
$$4^4 = 256$$

(D)

$$5^1 = 5$$
$$5^2 = 25$$
$$5^3 = 125$$
$$5^4 = 625$$

(E)

$$6^1 = 6$$
$$6^2 = 36$$
$$6^3 = 216$$

(F)

This procedure is further illustrated in the following problems:

1. $\sqrt{8} = \sqrt{4 \cdot 2}$
   $= \sqrt{4} \cdot \sqrt{2}$
   $= 2\sqrt{2}$

2. $\sqrt{72} = \sqrt{36 \cdot 2}$
   $= \sqrt{36} \cdot \sqrt{2}$
   $= 6\sqrt{2}$

By reference to the perfect fourth powers in table 7-1, we may simplify a radical such as $\sqrt[4]{405}$. Noting that 405 has the perfect fourth power 81 as a factor, we have the following:

$$\sqrt[4]{405} = \sqrt[4]{81 \cdot 5}$$
$$= \sqrt[4]{81} \cdot \sqrt[4]{5}$$
$$= 3\ (\sqrt[4]{5})$$

As was shown with fractional exponents, taking a root is equivalent to dividing the exponent of a power by the index of the root. If a factor of the radicand has an exponent that is not a multiple of the index of the root, the factor may be separated so that one exponent is divisible by the index, as in

$$\sqrt{3^7} = \sqrt{3^6 \cdot 3} = 3^{6/2} \cdot 3^{1/2} = 3^3 \cdot \sqrt{3} = 27\sqrt{3}$$

Consider also

$$\sqrt{2^3 \cdot 3^7 \cdot 5} = \sqrt{2^2 \cdot 2 \cdot 3^6 \cdot 3 \cdot 5}$$
$$= 2 \cdot 3^3\ (\sqrt{2 \cdot 3 \cdot 5})$$
$$= 54\ \sqrt{30}$$

If the radicand is a large number, the perfect powers that are factors are not always obvious. In such a case the radicand can be separated into prime factors. For example,

$$\sqrt{8,820} = \sqrt{2^2 \cdot 3^2 \cdot 5 \cdot 7^2}$$
$$= 2 \cdot 3 \cdot 7\sqrt{5}$$
$$= 42\ \sqrt{5}$$

Practice problems. Simplify the radicals and reduce to lowest terms:

1. $\dfrac{\sqrt{3} \cdot \sqrt{15}}{\sqrt{5}}$
  2. $\dfrac{\sqrt[3]{81}}{\sqrt[3]{27}}$

3. $\dfrac{18(\sqrt[3]{30})}{3(\sqrt[3]{10})}$ 4. $\dfrac{\sqrt{8,820}}{\sqrt{180}}$

Answers:

1. 3   3. $6(\sqrt[3]{3})$

2. $\sqrt[3]{3}$ 4. 7

RATIONAL AND IRRATIONAL NUMBERS

Real and imaginary numbers make up the number system of algebra. Imaginary numbers are discussed in chapter 15 of this course. Real numbers are either rational or irrational. The word RATIONAL comes from the word "ratio." A number is rational if it can be expressed as the quotient, or ratio, of two whole numbers. Rational numbers include fractions like 2/7, whole numbers, and radicals if the radical sign is removable.

Any whole number is rational. Its denominator is 1. For instance, 8 equals $\dfrac{8}{1}$, which is the quotient of two integers. A number like $\sqrt{16}$ is rational, since it can be expressed as the quotient of two integers in the form $\dfrac{4}{1}$. The following are also examples of rational numbers:

1. $\sqrt{\dfrac{25}{9}}$, which equals $\dfrac{5}{3}$

2.   -6, which equals $\dfrac{-6}{1}$

3. $5\dfrac{2}{7}$, which equals $\dfrac{37}{7}$

Any rational number can be expressed as the quotient of two integers in many ways. For example,

$$7 = \frac{7}{1} = \frac{14}{2} = \frac{21}{3} \cdots$$

An IRRATIONAL number is a real number that cannot be expressed as the ratio of two integers. The numbers $\sqrt{3}$, $5\sqrt{2}$, $\sqrt{7}$, $\frac{3}{8}\sqrt{20}$, and $\frac{2}{\sqrt{5}}$ are examples of irrational numbers.

Rationalizing Denominators

Expressions such as $\frac{7}{\sqrt{2}}$ and $\frac{\sqrt{2}}{5\sqrt{3}}$ have irrational numbers in the denominator. If the denominators are changed immediately to decimals, as in

$$\frac{7}{\sqrt{2}} = \frac{7}{1.4142}$$

the process of evaluating a fraction becomes an exercise in long division. Such a fraction can be evaluated quickly by first changing the denominator to a rational number. Converting a fraction with an irrational number in its denominator to an equivalent fraction with a rational number in the denominator is called RATIONALIZING THE DENOMINATOR.

Multiplying a fraction by 1 leaves the value of the fraction unchanged. Since any number divided by itself equals 1, it follows, for example, that

$$\frac{\sqrt{2}}{\sqrt{2}} = 1$$

If the numerator and denominator of $\frac{7}{\sqrt{2}}$ are each multiplied by $\sqrt{2}$, another fraction having the same value is obtained. The result is

$$\frac{7}{\sqrt{2}} = \frac{7}{\sqrt{2}} \cdot \frac{\sqrt{2}}{\sqrt{2}} = \frac{7\sqrt{2}}{2}$$

The denominator of the new equivalent fraction is 2, which is rational. The decimal value of the fraction is

$$\frac{7\sqrt{2}}{2} = \frac{7(1.4142)}{2} = 7(0.7071) = 4.9497$$

To rationalize the denominator in $\frac{\sqrt{2}}{5\sqrt{3}}$ we multiply the numerator and denominator by $\sqrt{3}$. We get

$$\frac{\sqrt{2}}{5\sqrt{3}} = \frac{\sqrt{2}}{5\sqrt{3}} \cdot \frac{\sqrt{3}}{\sqrt{3}} = \frac{\sqrt{6}}{5(3)} = \frac{\sqrt{6}}{15} \text{ or } \frac{1}{15}\sqrt{6}$$

Practice problems. Rationalize the denominator in each of the following:

1. $\frac{6}{\sqrt{2}}$

2. $\frac{\sqrt{5}}{\sqrt{3}}$

3. $\frac{2}{\sqrt{6}}$

4. $\frac{6}{\sqrt{y}}$

226

Answers:

1. $3\sqrt{2}$

2. $\dfrac{\sqrt{15}}{3}$

3. $\dfrac{\sqrt{6}}{3}$

4. $\dfrac{6\sqrt{y}}{y}$

EVALUATING RADICALS

Any radical expression has a decimal equivalent which may be exact if the radicand is a rational number. If the radicand is not rational, the root may be expressed as a decimal approximation, but it can never be exact. A procedure similar to long division may be used for calculating square root and cube root, and higher roots may be calculated by means of methods based on logarithms and higher mathematics. Tables of powers and roots have been calculated for use in those scientific fields in which it is frequently necessary to work with roots.

SQUARE ROOT PROCESS

The arithmetic process for calculation of square root is outlined in the following paragraphs:

1. Begin at the decimal point and mark the number off into groups of two digits each, moving both to the right and to the left from the decimal point. This may leave an odd digit at the right-hand or left-hand end of the number, or both. For example, suppose that the number whose square root we seek is 9025. The number marked off as specified would be as follows:

$$\sqrt{90'25}.$$

2. Find the greatest number whose square is contained in the left-hand group (90). This number is 9, since the square of 9 is 81. Write 9 above the first group. Square this number (9), place its square below the left-hand group, and subtract, as follows:

$$
\begin{array}{r}
9 \\[2pt]
\hline
\sqrt{90'25}. \\[2pt]
81 \\[2pt]
\hline
9\ 25
\end{array}
$$

Bring down the next group (25) and place it beside the 9, as shown. This is the new dividend (925).

3. Multiply the first digit in the root (9) by 20, obtaining 180 as a trial divisor. This trial divisor is contained in the new dividend (925) five times; thus the second digit of the root appears to be 5. However, this number must be added to the trial divisor to obtain a "true divisor." If the true divisor is then too large to use with the second quotient digit, this digit must be reduced by 1. The procedure for step 3 is illustrated as follows:

$$
\begin{array}{r r}
 & 9\quad 5. \\[2pt]
 & \overline{\sqrt{90'25}.} \\[2pt]
 & 81 \\[2pt]
\cancel{180} & \overline{9\ 25} \\[2pt]
185 & 9\ 25 \\[2pt]
 & \overline{0\ 00}
\end{array}
$$

The number 180, resulting from the multiplication of 9 by 20, is written as a trial divisor beside the new dividend (925), as shown. The quotient digit (5) is then recorded and the trial divisor is adjusted, becoming 185. The trial quotient (180) is crossed out.

4. The true divisor (185) is multiplied by the second digit (5) and the product is placed below the new dividend (925). This step is shown in the illustration for step 3. When the product in step 4 is subtracted from the new dividend, the difference is 0; thus, in this example, the root is exact.

5. In some problems, the difference is not 0 after all of the digits of the original number have been used to form new dividends. Such problems may be carried further by adding 0's on the right-hand end of the original number, just as in normal long division. However, in the square root process the 0 s must be added and used in groups of 2.

Practice problems. Find the square root of each of the following numbers:

1. 9.61   2. 123.21   3. 0.0025

Answers:

1. 3.1   2. 11.1   3. 0.05

TABLES OF ROOTS

The decimal values of square roots and cube roots of numbers with as many as 3 or 4 digits can be found from tables. The table in appendix I of this course gives the square roots and cube roots of numbers from 1 to 100. Most of the values given in such tables are approximate numbers which have been rounded off.

For example, the fourth column in appendix I shows that $\sqrt{72} = 8.4853$, to 4 decimal places. By shifting the decimal point we can obtain other square roots. A shift of two places in the decimal point in the radicand corresponds to a shift of one place in the same direction in the square root.

The following examples show the effect, as reflected in the square root, of shifting the location of the decimal point in the number whose square root we seek:

$$\sqrt{72} = 8.4853$$
$$\sqrt{0.72} = 0.84853$$
$$\sqrt{0.0072} = 0.084853$$
$$\sqrt{7,200} = 84.853$$

Cube Root

The fifth column in appendix I shows that the cube root of 72 is 4.1602. By shifting the decimal point we immediately have the cube roots of certain other numbers involving the same digits. A shift of three places in the decimal point in the radicand corresponds to a shift of one place in the same direction in the cube root.

Compare the following examples:

$$\sqrt[3]{72} = 4.1602$$

$$\sqrt[3]{0.072} = 0.41602$$

$$\sqrt[3]{72,000} = 41.602$$

Many irrational numbers in their simplified forms involve $\sqrt{2}$ and $\sqrt{3}$. Since these radicals occur often, it is convenient to remember their decimal equivalents as follows:

$$\sqrt{2} = 1.4142 \text{ and } \sqrt{3} = 1.7321$$

Thus any irrational numbers that do not contain any radicals other than $\sqrt{2}$ or $\sqrt{3}$ can be converted to decimal forms quickly without referring to tables.

For example consider

$$\sqrt{72} = 6\sqrt{2} = 6(1.4142) = 8.485$$

$$\sqrt{27} = 3\sqrt{3} = 3(1.7321) = 5.196$$

Keep in mind that the decimal equivalents of $\sqrt{2}$ and $\sqrt{3}$ as used in the foregoing examples are not exact numbers and the results obtained with them are approximate in the fourth decimal place.

CHAPTER 8

LOGARITHMS

Logarithms represent a specialized use of exponents. By means of logarithms, computation with large masses of data can be greatly simplified. For example, when logarithms are used, the process of multiplication is replaced by simple addition and division is replaced by subtraction. Raising to a power by means of logarithms is done in a single multiplication, and extracting a root reduces to simple division.

DEFINITIONS

In the expression $2^3 = 8$, the number 2 is the base (not to be confused with the base of the number system), and 3 is the exponent which must be used with the base to produce the number 8. The exponent 3 is the logarithm of 8 when the base is 2. This relationship is usually stated as follows: The logarithm of 8 to the base 2 is 3. In general, the logarithm of a number N with respect to a given base is the exponent which must be used with the base to produce N. Table 8-1 illustrates this.

Table 8-1 shows that the logarithmic relationship may be expressed equally well in either of two forms; these are the exponential form and the logarithmic form. Observe, in table 8-1, that the base of a logarithmic expression is indicated by placing a subscript just below and to the right of the abbreviation "log." Observe also that the word "logarithm" is abbreviated without using a period.

Table 8-1.—Logarithms with various bases.

Exponential form	Logarithmic form
$2^3 = 8$	$\log_2 8 = 3$
$4^2 = 16$	$\log_4 16 = 2$
$5^0 = 1$	$\log_5 1 = 0$
$27^{2/3} = 9$	$\log_{27} 9 = 2/3$

The equivalency of the logarithmic and exponential forms may be used to restate the fundamental definition of logarithms in its most useful form, as follows:

$$b^x = N \text{ implies that } \log_b N = x$$

In words, this definition is stated as follows: If the base b raised to the x power equals N, then x is the logarithm of the number N to the base b.

One of the many uses of logarithms may be shown by an example in which the base is 2. Table 8-2 shows the powers of 2 from 0 through 20. Suppose that we wish to use logarithms to multiply the numbers 512 and 256, as follows:

From table 8-2, $512 = 2^9$

$256 = 2^8$

Then $512 \times 256 = 2^9 \times 2^8$

$= 2^{17}$

and from the table again $2^{17} = 131072$

It is seen that the problem of multiplication is reduced to the simple addition of the exponents 9 and 8 and finding the corresponding power in the table.

Table 8-2 (A) shows the base 2 in the exponential form with its corresponding powers. The actual computation in logarithmic work does not require that we record the exponential form. All that is required is that we add the appropriate exponents and have available a table in which we can look up the number corresponding to the new exponent after adding. Therefore, table 8-2 (B) is adequate for our purpose. Solving the foregoing example by this table, we have the following:

$$\log_2 512 = 9$$

$$\log_2 256 = 8$$

$$\log_2 \text{ of the product} = 17$$

Therefore, the number we seek is the one in the table whose logarithm is 17. This number is 131,072. In this example, we found the exponents directly, added them since this was a multiplication problem, and located the corresponding power. This avoided the unnecessary step of writing the base 2 each time.

Table 8-2.—Exponential and logarithmic
tables for the base 2.

(A) Powers of 2 from 0 through 20		(B) Logarithms for the base 2 and corresponding powers	
		Log	Number
$2^0 =$	1	0	1
$2^1 =$	2	1	2
$2^2 =$	4	2	4
$2^3 =$	8	3	8
$2^4 =$	16	4	16
$2^5 =$	32	5	32
$2^6 =$	64	6	64
$2^7 =$	128	7	128
$2^8 =$	256	8	256
$2^9 =$	512	9	512
$2^{10} =$	1024	10	1024
$2^{11} =$	2048	11	2048
$2^{12} =$	4096	12	4096
$2^{13} =$	8192	13	8192
$2^{14} =$	16384	14	16384
$2^{15} =$	32768	15	32768
$2^{16} =$	65536	16	65536
$2^{17} =$	131072	17	131072
$2^{18} =$	262144	18	262144
$2^{19} =$	524288	19	524288
$2^{20} =$	1048576	20	1048576

Practice problems. Use the logarithms in table 8-2 to perform the following multiplication:

1. 64 x 128

2. 1,024 x 256

3. 128 x 4,096

4. 512 x 2,048

Answers:

1. 8,192

2. 262,144

3. 524,288

4. 1,048,576

NATURAL AND COMMON LOGARITHMS

Many natural phenomena, such as rates of growth and decay, are most easily described in terms of logarithmic or exponential formulas. Furthermore, the geometric patterns in which certain seeds grow (for example, sunflower seeds) is a logarithmic spiral. These facts explain the name "natural logarithms." Natural logarithms use the base e, which is an irrational number approximately equal to 2.71828. This system is sometimes called the Napierian system of logarithms, in honor of John Napier, who is credited with the invention of logarithms.

To distinguish natural logarithms from other logarithmic systems the abbreviation, ln, is sometimes used. When ln appears, the base is understood to be e and need not be shown. For example, either $\log_e 45$ or ln 45 signifies the natural logarithm of 45.

COMMON LOGARITHMS

As has been shown in preceding paragraphs, any number may be used as a base for a system of logarithms. The selection of a base is a matter of convenience. Briggs in 1617 found that base 10 possessed many advantages not obtainable in ordinary calculations with other bases. The selection of 10 as a base proved so satisfactory that today it is used almost exclusively for ordinary calculations. Logarithms with 10 as a base are therefore called COMMON LOGARITHMS.

When 10 is used as a base, it is not necessary to indicate it in writing logarithms. For example,

$$\log \; 100 = 2$$

is understood to mean the same as

$$\log_{10} \; 100 = 2$$

If the base is other than 10, it must be specified by the use of a subscript to the right and below the abbreviation "log." As noted in the foregoing discussion of natural logarithms, the use of the distinctive abbreviation "ln" eliminates the need for a subscript when the base is e.

It is relatively easy to convert common logarithms to natural logarithms or vice versa, if necessary. It should be noted further that each system has its peculiar advantages, but for most everyday work, the common system is more often used. A simple relation connects the two systems. If the common logarithm of a number can be found, multiplying by 2.3026 gives the natural logarithm of the number. For example,

$$\log\ 1.60 = 0.2041$$

$$\ln\ 1.60 = 2.3026 \times 0.2041$$

$$= 0.4700$$

Thus the natural logarithm of 1.60 is 0.4700, correct to four significant digits.

Conversely, multiplying the natural logarithm by 0.4343 gives the common logarithm of a number. As might be expected, the conversion factor 0.4343 is the reciprocal of 2.3026. This is shown as follows:

$$\frac{1}{2.3026} = 0.4343$$

Positive Integral Logarithms

The derivation of positive whole logarithms is readily apparent. For example, we see in table 8-3 (B) that the logarithm of 10 is 1. The number 1 is simply the exponent of the base 10 which yields 10. This is shown in table 8-3 (A) opposite the logarithmic equation. Similarly,

$$10^0 = 1 \ldots\ldots\ldots \log\ 1 = 0$$

$$10^2 = 100 \ldots\ldots \log\ 100 = 2$$

$$10^3 = 1,000 \ldots\ldots \log\ 1,000 = 3$$

$$10^4 = 10,000 \ldots\ldots \log\ 10,000 = 4$$

Positive Fractional Logarithms

Referring to table 8-3, notice that the logarithm of 1 is 0 and the logarithm of 10 is 1. Therefore, the logarithm of a number between 1 and 10 is between 0 and 1. An easy way to

Table 8-3.—Exponential and corresponding logarithmic notations using base 10.

10^{-4}	$= \dfrac{1}{10^4}$	$=$	0.0001	log	0.0001	$= -4$
10^{-3}	$= \dfrac{1}{10^3}$	$=$	0.001	log	0.001	$= -3$
10^{-2}	$= \dfrac{1}{10^2}$	$=$	0.01	log	0.01	$= -2$
10^{-1}	$= \dfrac{1}{10}$	$=$	0.1	log	0.1	$= -1$
$10^{-1/2}$	$= \dfrac{1}{\sqrt{10}} = \dfrac{\sqrt{10}}{10} =$		0.31623	log	0.31623	$= -0.5$
						$= 0.5 - 1$
10^{0}		$=$	1	log	1	$= 0$
$10^{1/2}$	$= \sqrt{10}$	$=$	3.1623	log	3.1623	$= 0.5$
10^{1}		$=$	10	log	10	$= 1$
$10^{3/2}$	$= 10\sqrt{10}$	$=$	31.623	log	31.623	$= 1.5$
10^{2}		$=$	100	log	100	$= 2$
$10^{5/2}$	$= 10^2 (\sqrt{10})$	$=$	316.23	log	316.23	$= 2.5$
10^{3}		$=$	$1,000$	log	$1,000$	$= 3$
$10^{7/2}$	$= 10^3 (\sqrt{10})$	$=$	3162.3	log	3162.3	$= 3.5$
10^{4}		$=$	$10,000$	log	$10,000$	$= 4$

verify this is to consider some numbers between 1 and 10 which are powers of 10; the exponent in each case will then be the logarithm we seek. Of course, the only powers of 10 which produce numbers between 1 and 10 are fractional powers.

EXAMPLE: $10^{1/2}$ = 3.1623 (approximately)

$10^{0.5}$ = 3.1623

Therefore, log 3.1623 = 0.5

Other examples are shown in the table for

$10^{3/2}$, $10^{5/2}$, and $10^{7/2}$. Notice that the number that represents $10^{3/2}$, 31.623, logically enough lies between the numbers representing 10^1 and 10^2—that is, between 10 and 100. Notice also that $10^{5/2}$ appears between 10^2 and 10^3, and $10^{7/2}$ lies between 10^3 and 10^4.

Negative Logarithms

Table 8-3 shows that negative powers of 10 may be fitted into the system of logarithms. We recall that 10^{-1} means $\frac{1}{10}$, or the decimal fraction, 0.1. What is the logarithm of 0.1?

SOLUTION: $10^{-1} = 0.1$; log 0.1 = -1

Likewise $10^{-2} = 0.01$; log 0.01 = -2

Negative Fractional Logarithms

Notice in table 8-3 that negative fractional exponents present no new problem in logarithmic notation. For example, $10^{-1/2}$ means $\frac{1}{\sqrt{10}}$.

$$\frac{1}{\sqrt{10}} = \frac{\sqrt{10}}{10} = 0.31623$$

What is the logarithm of 0.31623?

SOLUTION:

$$10^{-1/2} = 0.31623; \log 0.31623 = -\frac{1}{2}$$

$$= -0.5$$

Table 8-3 shows logarithms for numbers ranging from 0.0001 to 10,000. Notice that there are only 8 integral logarithms in the entire range. Excluding zero logarithms, the logarithms for all other numbers in the range are fractional or contain a fractional part. By the year 1628, logarithms for all integers from 1 to 100,000 had been computed. Practically all of these logarithms contain a fractional part. It should be remembered that finding the logarithm of a number is nothing more than expressing the number as a power of 10. Table 8-4 shows the numbers 1 through 10 expressed as powers of 10. Most of the exponents which comprise logarithms are found by methods beyond the scope of this text. However, it is not necessary to know the process used to obtain logarithms in order to make use of them.

Table 8-4.—The numbers 1 through 10 expressed as powers of 10.

$1 = 10^0$	$6 = 10^{0.77815}$
$2 = 10^{0.30103}$	$7 = 10^{0.84510}$
$3 = 10^{0.47712}$	$8 = 10^{0.90309}$
$4 = 10^{0.60206}$	$9 = 10^{0.95424}$
$5 = 10^{0.69897}$	$10 = 10^1$

COMPONENTS OF LOGARITHMS

The fractional part of a logarithm is usually written as a decimal. The whole number part of a logarithm and the decimal part have been given separate names because each plays a special part in relation to the number which the logarithm represents. The whole number part of a logarithm is called the CHARACTERISTIC. This part of the logarithm shows the position of the decimal point in the associated number. The decimal part of a logarithm is called the MANTISSA.

For a particular sequence of digits making up a number, the mantissa of a common logarithm is always the same regardless of the position of the decimal point in that number. For example, log 5270 = 3.72181; the mantissa is 0.72181 and the characteristic is 3.

CHARACTERISTIC

The characteristic of a common logarithm shows the position of the decimal point in the associated number. The characteristic for a given number may be determined by inspection. It will be remembered that a common logarithm is simply an exponent of the base 10.

When we write log 360 = 2.55630, we understand this to mean $10^{2.55630} = 360$. We know that the number is 360 and not 36 or 3,600 because the characteristic is 2. We know 10^1 is 10, 10^2 is 100, and 10^3 is 1,000. Therefore, the number whose value is $10^{2.55630}$ must lie between 100 and 1,000 and of course any number in that range has 3 digits.

Suppose the characteristic had been 1: where would the decimal point in the number be placed? Since 10^1 is 10 and 10^2 is 100, any number whose logarithm is between 1 and 2 must lie between 10 and 100 and will have 2 digits. Notice how the position of the decimal point changes with the value of the characteristic in the following examples:

$$\log 36{,}000 = 4.55630$$

$$\log 3{,}600 = 3.55630$$

$$\log 360 = 2.55630$$

$$\log 36 = 1.55630$$

$$\log 3.6 = 0.55630$$

Note that it is only the characteristic that changes when the decimal point is moved. An advantage of using the base 10 is thus revealed: If the characteristic is known, the decimal point may easily be placed. If the number is known, the characteristic may be determined by inspection; that is, by observing the location of the decimal point.

Although an understanding of the relation of the characteristic to the powers of 10 is necessary for thorough comprehension of logarithms, the characteristic may be determined mechanically by application of the following rules:

1. For a number greater than 1, the characteristic is positive and is one less than the number of digits to the left of the decimal point in the number.

2. For a positive number less than 1, the characteristic is negative and has an absolute value one more than the number of zeros between the decimal point and the first nonzero

Table 8-5.—Positive and negative characteristics.

Number	Power of 10	Digits in number to the left of decimal point	Characteristic
	Between:		
134	10^2 and 10^3	3	2
13.4	10^1 and 10^2	2	1
1.34	10^0 and 10^1	1	0
		Zeros between decimal point and first non-zero digit	
0.134	10^{-1} and 10^0	0	-1
0.0134	10^{-2} and 10^{-1}	1	-2
0.00134	10^{-3} and 10^{-2}	2	-3

digit of the number.

Table 8-5 contains examples of each type of characteristic.

Practice problems. In problems 1 through 4, write the characteristic of the logarithm for each number. In 5 through 8, place the decimal point in each number as indicated by the characteristic (c) given for each.

1. 4,321   2. 1.23   3. 0.05   4. 12

5. 123; c = 4   6. 8,210; c = 0

7. 8; c = -1   8. 321; c = -2

Answers:

1. 3   2. 0   3. -2   4. 1

5. 12,300   6. 8.210   7. 0.8   8. 0.0321

Negative Characteristics

When a characteristic is negative, such as -2, we do not carry out the subtraction, since this would involve a negative mantissa. There are several ways of indicating a negative characteristic. Mantissas as presented in the table in the appendix are always positive and the sign of the characteristic is indicated separately. For example, where $\log 0.023 = \bar{2}.36173$, the bar over the 2 indicates that only the characteristic is negative—that is, the logarithm is -2 + 0.36173.

Another way to show the negative characteristic is to place it after the mantissa. In this case we write 0.36173-2.

A third method, which is used where possible throughout this chapter, is to add a certain quantity to the characteristic and to subtract the same quantity to the right of the mantissa. In the case of the example, we may write:

$$\overline{2}.36173$$

$$\underline{10 \qquad -10}$$

$$8.36173\text{-}10$$

In this way the value of the logarithm remains the same but we now have a positive characteristic as well as a positive mantissa.

MANTISSA

The mantissa is the decimal part of a logarithm. Tables of logarithms usually contain only mantissas since the characteristic can be readily determined as explained previously. Table 8-6 shows the characteristic, mantissa, and logarithm for several positions of the decimal point using the sequence of digits 4, 5, 6. It will be noted that the mantissa remains the same for that particular sequence of digits, regardless of the position of the decimal point.

Appendix I of this training course is a table which includes the logarithms of numbers from 1 to 100. For our present purpose in using this table, we are concerned only with the first and sixth columns.

The first column contains the number and the sixth column contains its logarithm. For example, if it is desired to find the logarithm of 45, we would find the number 45 in the first

Table 8-6.—Effect of changes in the
location of the decimal point.

Number	Charac-teristic	Mantissa	Logarithm
45,600	4	0.6590	4.6590
4,560	3	0.6590	3.6590
456	2	0.6590	2.6590
45.6	1	0.6590	1.6590
4.56	0	0.6590	0.6590
0.456	-1	0.6590	0.6590-1
0.0456	-2	0.6590	0.6590-2
0.00456	-3	0.6590	0.6590-3

column, look horizontally across the page to column 6 and read the logarithm, 1.65321. A glance down the logarithm column will reveal that the logarithms increase in value as the numbers increase in value.

It must be noted in this particular table that both the mantissa and the characteristic are given for the number in the first column. This is simply an additional aid, since the characteristic can easily be determined by inspection.

Suppose that we wish to use the table of Appendix I to find the logarithm of a number not shown in the "number" column. By recalling that the mantissa does not change when the decimal point moves, we may be able to deter-

247

mine the desired logarithm. For example, the number 450 does not appear in the number column of the table. However, the number 45 has the same mantissa as 450; the only difference between the two logs is in their characteristics. Thus the logarithm of 450 is 2.65321.

Practice problems. Find the logarithms of the following numbers:

1. 64   2. 98   3. 6400   4. 9.8

Answers:

1. 1.80618   2. 1.99123

3. 3.80618   4. 0.99123

CHAPTER 9

FUNDAMENTALS OF ALGEBRA

The numbers and operating rules of arithmetic form a part of a very important branch of mathematics called ALGEBRA.

Algebra extends the concepts of arithmetic so that it is possible to generalize the rules for operating with numbers and use these rules in manipulating symbols other than numbers. It does not involve an abrupt change into a distinctly new field, but rather provides a smooth transition into many branches of mathematics with a continuation of knowledge already gained in basic arithmetic.

The idea of expressing quantities in a general way, rather than in the specific terms of arithmetic, is fairly common. A typical example is the formula for the perimeter of a rectangle, $P = 2L + 2W$, in which the letter P represents perimeter, L represents length, and W represents width. It should be understood that $2L = 2(L)$ and $2W = 2(W)$. If the L and the W were numbers, parentheses or some other multiplication sign would be necessary, but the meaning of a term such as 2L is clear without additional signs or symbols.

All formulas are algebraic expressions, although they are not always identified as such. The letters used in algebraic expressions are often referred to as LITERAL NUMBERS (literal implies "letteral").

Another typical use of literal numbers is in the statement of mathematical laws of operation. For example, the commutative, associative, and distributive laws, introduced in chapter 3 with respect to arithmetic, may be restated in general terms by the use of algebraic symbols.

COMMUTATIVE LAWS

The word "commutative" is defined in chapter 3. Remember that the commutative laws refer to those situations in which the factors and terms of an expression are rearranged in a different order.

Addition

The algebraic form of the commutative law for addition is as follows:

$$a + b + c = a + c + b = c + b + a$$

In words, this law states that the sum of two or more addends is the same regardless of the order in which the addends are arranged.

The arithmetic example in chapter 3 shows only one specific numerical combination in which the law holds true. In the algebraic example, a, b, and c represent any numbers we choose, thus giving a broad inclusive example of the rule. (Note that once a value is selected for a literal number, that value remains the same wherever the letter appears in that particular example or problem. Thus, if we give a the value of 12, in the example just given, a's value is 12 wherever it appears.)

Multiplication

The algebraic form of the commutative law for multiplication is as follows:

abc = acb = cba

In words, this law states that the product of two or more factors is the same regardless of the order in which the factors are arranged.

ASSOCIATIVE LAWS

The associative laws of addition and multiplication refer to the grouping (association) of terms and factors in a mathematical expression.

Addition

The algebraic form of the associative law for addition is as follows:

$$a + b + c = (a + b) + c = a + (b + c)$$

In words, this law states that the sum of three or more addends is the same regardless of the manner in which the addends are grouped.

Multiplication

The algebraic form of the associative law for multiplication is as follows:

$$a \cdot b \cdot c = (a \cdot b) \cdot c = a \cdot (b \cdot c)$$

In words, this law states that the product of

three or more factors is the same regardless of the manner in which the factors are grouped.

DISTRIBUTIVE LAW

The distributive law refers to the distribution of factors among the terms of an additive expression. The algebraic form of this law is as follows:

$$a(b + c + d) = ab + ac + ad$$

In words, this law may be stated as follows: If the sum of two or more quantities is multiplied by a third quantity, the product is found by applying the multiplier to each of the original quantities separately and summing the resulting expressions.

ALGEBRAIC SUMS

The word "sum" has been used several times in this discussion, and it is important to realize the full implication where algebra is concerned. Since a literal number may represent either a positive or a negative quantity, a sum of several literal numbers is always understood to be an ALGEBRAIC SUM. That is, it is the sum that results when the algebraic signs of all the addends are taken into consideration.

The following problems illustrate the procedure for finding an algebraic sum:

Let $a = 3$, $b = -2$, and $c = 4$.

Then $a + b + c = (3) + (-2) + (4)$

$$= 5$$

Also, $a - b - c = a + (-b) + (-c)$
$$= 3 + (+2) + (-4)$$
$$= 1$$

The second problem shows that every expression containing two or more terms to be combined by addition and subtraction may be rewritten as an algebraic sum, all negative signs being considered as belonging to specific terms and all operational signs being positive.

It should be noted, in relation to this subject, that the laws of signs for algebra are the same as those for arithmetic.

ALGEBRAIC EXPRESSIONS

An algebraic expression is made up of the signs and symbols of algebra. These symbols include the Arabic numerals, literal numbers, the signs of operation, and so forth. Such an expression represents one number or one quantity. Thus, just as the sum of 4 and 2 is one quantity, that is, 6, the sum of c and d is one quantity, that is, $c + d$. Likewise $\frac{a}{b}$, $\sqrt{b}$, ab, $a - b$, and so forth, are algebraic expressions each of which represents one quantity or number.

Longer expressions may be formed by combinations of the various signs of operation and the other algebraic symbols, but no matter how complex such expressions are they still represent one number. Thus the algebraic expression $\frac{-a + \sqrt{2a + b}}{6} - c$ is one number

The arithmetic value of any algebraic expression depends on the values assigned to the literal numbers. For example, in the expression $2x^2 - 3ay$, if $x = -3$, $a = 5$, and $y = 1$, then we have the following:

$$2x^2 - 3ay = 2(-3)^2 -3(5)(1)$$
$$= 2(9) - 15 = 18 - 15 = 3$$

Notice that the exponent is an expression such as $2x^2$ applies only to the x. If it is desired to indicate the square of 2x, rather than 2 times the square of x, then parentheses are used and the expression becomes $(2x)^2$.

Practice problems. Evaluate the following algebraic expressions when $a = 4$, $b = 2$, $c = 3$, $x = 7$, and $y = 5$. Remember, the order of operation is multiplication, division, addition, and subtraction.

1. $3x + 7y - c$

2. $xy - 4a^2$

3. $\dfrac{ax}{b} + y$

4. $c + \dfrac{ay^2}{b}$

Answers:

1. 53

2. -29

3. 19

4. 53

TERMS AND COEFFICIENTS

The terms of an algebraic expression are the parts of the expression that are connected by plus and minus signs. In the expression 3abx + cy - k, for example, 3abx, cy, and k are the terms of the expression.

An expression containing only one term, such as 3ab, is called a monomial (mono means one). A binominal contains two terms; for example, 2r + by. A trinomial consists of three terms. Any expression containing two or more terms may also be called by the general name, polynomial (poly means many). Usually special names are not given to polynomials of more than three times. The expression $x^3 - 3x^2 + 7x + 1$ is a polynomial of four terms. The trinomial $x^2 + 2x + 1$ is an example of a polynomial which has a special name.

Practice problems. Identify each of the following expressions as a monomial, binomial, trinomial, or polynomial. (Some expressions may have two names.)

1. x

2. 3y + a + b

3. abx

4. 4 + 2b + y + z

5. $3y^2 + 4$

6. $\frac{2y}{6} + 1$

Answers:

1. Monomial

2. Trinomial
(also polynomial)

3. Monomial

4. Polynomial

5. Binomial
(also polynomial)

6. Binomial
(also polynomial)

In general, a COEFFICIENT of a term is any factor or group of factors of a term by which the remainder of the term is to be multiplied. Thus in the term 2axy, 2ax is the coefficient of y, 2a is the coefficient of xy, and 2 is the coefficient of axy. The word "coefficient" is usually used in reference to that factor which is expressed in Arabic numerals. This factor is sometimes called the NUMERICAL COEFFICIENT. The numerical coefficient is customarily written as the first factor of the term. In 4x, 4 is the numerical coefficient, or simply the coefficient, of x. Likewise, in $24xy^2$, 24 is the coefficient of xy^2 and in 16(a + b), 16 is the coefficient of (a + b). When no numerical coefficient is written it is understood to be 1. Thus in the term xy, the coefficient is 1.

COMBINING TERMS

When arithmetic numbers are connected by plus and minus signs, they can always be combined into one number. Thus,

$$5 - 7\frac{1}{2} + 8 = 5\frac{1}{2}$$

Here three numbers are added algebraically (with due regard for sign) to give one number. The terms have been combined into one term.

Terms containing literal numbers can be combined only if their literal parts are the same. Terms containing literal factors in which the same letters are raised to the same power are called like terms. For example, 3y and 2y are like terms since the literal parts are the same. Like terms are added by adding the coefficients of the like parts. Thus, 3y + 2y = 5y just as 3 bolts + 2 bolts = 5 bolts. Also

$3a^2b$ and a^2b are like; $3a^2b + a^2b = 4a^2b$ and $3a^2b - a^2b = 2a^2b$. The numbers ay and by are like terms with respect to y. Their sum could be indicated in two ways: $ay + by$ or $(a + b)y$. The latter may be explained by comparing the terms to denominate numbers. For instance, a bolts + b bolts = (a + b) bolts.

Like terms are added or subtracted by adding or subtracting the numerical coefficients and placing the result in front of the literal factor, as in the following examples:

$$7x^2 - 5x^2 = (7 - 5)x^2 = 2x^2$$
$$5b^2x - 3ay^2 - 8b^2x + 10ay^2 = -3b^2x + 7ay^2$$

Dissimilar or unlike terms in an algebraic expression cannot be combined when numerical values have not been assigned to the literal factors. For example, $-5x^2 + 3xy - 8y^2$ contains three dissimilar terms. This expression cannot be further simplified by combining terms through addition or subtraction. The expression may be rearranged as $x(3y - 5x) - 8y^2$ or $y(3x - 8y) - 5x^2$, but such a rearrangement is not actually a simplification.

Practice problems. Combine like terms in the following expression:

1. $2a + 4a$

2. $y + y^2 + 2y$

3. $4\dfrac{ay}{c} - \dfrac{ay}{c}$

4. $2ay^2 - ay^2$

5. $bx^2 + 2bx^2$

6. $2y + y^2$

Answers:

1. 6a

2. $y^2 + 3y$

3. $3\dfrac{ay}{c}$

4. ay^2

5. $3bx^2$

6. $2y + y^2$

SYMBOLS OF GROUPING

Often it is desired to group two or more terms to indicate that they are to be considered and treated as though they were one term even though there may be plus and minus signs between them. The symbols of grouping are parentheses ( ) (which we have already used), brackets [ ], braces { }, and the vinculum ___. The vinculum is sometimes called the "overscore." The fact that -7 + 2 - 5 is to be subtracted from 15, for example, could be indicated in any one of the following ways:

$$15 - (-7 + 2 - 5)$$
$$15 - [-7 + 2 - 5]$$
$$15 - \{-7 + 2 - 5\}$$
$$15 - \overline{-7 + 2 - 5}$$

Actually the vinculum is seldom used except in connection with a radical sign, such as in $\sqrt{a + b}$, or in a Boolean algebra expression. Boolean algebra is a specialized kind of symbolic notation which is discussed in Mathematics, Volume 3, NavPers 10073.

Parentheses are the most frequently used

symbols of grouping. When several symbols are needed to avoid confusion in grouping, parentheses usually comprise the innermost symbols, followed by brackets, and then by braces as the outermost symbols. This arrangement of grouping symbols is illustrated as follows:

$$2x - \{3y + [- 8 - 5y - (x - 4)]\}$$

REMOVING AND INSERTING GROUPING SYMBOLS

Discussed in the following paragraphs are various rules governing the removal and insertion of parentheses, brackets, braces, and the vinculum. Since the rules are the same for all grouping symbols, the discussion in terms of parentheses will serve as a basis for all.

Removing Parentheses

If parentheses are preceded by a minus sign, the entire quantity enclosed must be regarded as a subtrahend. This means that each term of the quantity in parentheses is subtracted from the expression preceding the minus sign. Accordingly, parentheses preceded by a minus sign can be removed, if the signs of all terms within the parentheses are changed.

This may be explained with an arithmetic example. We recall that to subtract one number from another, we change the sign of the subtrahend and proceed as in addition. To subtract -7 from 16, we change the sign of -7 and proceed as in addition, as follows:

$$16 - (-7) = 16 + 7$$
$$= 23$$

It is sometimes easier to see the result of changing signs in the subtrahend if the minus sign preceding the parentheses is regarded as a multiplier. Thus, the thought process in removing parentheses from an expression such as $- (4 - 3 + 2)$ would be as follows: Minus times plus is minus, so the first term of the expression with parentheses removed is $- 4$. (Remember that the 4 in the original expression is understood to be a +4, since it has no sign showing.) Minus times minus is plus, so the second term is +3. Minus times plus is minus, so the third term is -2. The result is $- 4 + 3 - 2$, which reduces to -3.

This same result can be reached just as easily, in an arithmetic expression, by combining the numbers within the parentheses before applying the negative sign which precedes the parentheses. However, in an algebraic expression with no like terms such combination is not possible. The following example shows how the rule for removal of parentheses is applied to algebraic expressions:

$$2a - (-4x + 3by) = 2a + 4x - 3by$$

Parentheses preceded by a plus sign can be removed without any other changes, as the following example shows:

$$2b + (a - b) = 2b + a - b = a + b$$

Many expressions contain more than one set of parentheses, brackets, and other symbols of grouping. In removing symbols of grouping, it is possible to proceed from the outside inward or from the inside outward. For the beginner, it is simpler to start on the inside and work toward the outside, collecting terms and simplifying as one proceeds. In the following example the inner grouping symbols are removed first:

$$2a - [x + (x - 3a) - (9a - 5x)]$$
$$= 2a - [x + x - 3a - 9a + 5x]$$
$$= 2a - [7x - 12a]$$
$$= 2a - 7x + 12a$$
$$= 14a - 7x$$

Enclosing Terms in Parentheses

When it is desired to enclose a group of terms in parentheses, the group of terms remains unchanged if the sign preceding the parentheses is positive. This is illustrated as follows:

$$3x - 2y + 7x - y = (3x - 2y) + (7x - y).$$

Note that this agrees with the rule for removing parentheses preceded by a plus sign.

If terms are enclosed within parentheses preceded by a minus sign, the signs of all the terms enclosed must be changed as in the following example:

$$3x - 2y + 7x - y = 3x - (2y - 7x + y)$$

261

Practice problems. In problems 1 through 4, remove the symbols of grouping and combine like terms. In problems 5 through 8, enclose the first two terms in parentheses preceded by a plus sign (understood) and the last two in parentheses preceded by a minus sign.

1. $6a - (4a - 3)$

2. $3x + [2x - 4y(6 - 4x)] + 2y - (3 - x + 3y)$

3. $-a + [-a - (2a + 3)] + 3$

4. $(7x - 3ay) - (4a - b) + 16$

5. $4a - 3b - 2c + 4d$

6. $-2 -3x +4y - z$

7. $x + 4y + 3z + 7$

8. $-4 + 2a - 6c + 3d$

Answers:

1. $2a + 3$

2. $6x + 16xy - 25y - 3$

3. $-4a$

4. $7x - 3ay - 4a + b + 16$

5. $(4a - 3b) - (2c - 4d)$

6. $(-2 -3x) - (-4y + z)$

7. $(x + 4y) - (-3z - 7)$

8. $(-4 + 2a) - (6c - 3d)$

EXPONENTS AND RADICALS

Exponents and radicals have the same meaning in algebra as they do in arithmetic. Thus, if n represents any number then $n^2 = n \cdot n$, $n^3 = n \cdot n \cdot n$, etc. By the same reasoning, n^m means that n is to be taken as a factor m times. That is, n^m is equal to $n \cdot n \cdot n \cdot \ldots$, with n appearing m times. The series of dots, called ellipsis (not to be confused with the geometric figure having a similar name, ellipse), represents continuation of the same pattern or the same symbol.

The rules of operation with exponents are also the same in algebra as in arithmetic. For example, $n^2 \cdot n^3 = n^{2+3} = n^5$. Some care is necessary to avoid confusion over an expression such as $3^2 \cdot 3^3$. In this example, n = 3 and the product desired is 3^5, not 9^5. In general, $3^a \cdot 3^b = 3^{a+b}$, and a similar result is reached whether the factor which acts as a base for the exponents is a number or a letter. Thus the general form can be expressed as follows:

$$n^a \cdot n^b = n^{a+b}$$

In words, the general rule for multiplication involving exponents is as follows: When multiplying terms whose literal factors are like, the exponents are added. This rule may be applied to problems involving division, if all expressions containing exponents in denominators are rewritten as expressions with negative exponents. For example, the fraction $\dfrac{x^2 y}{x y^2}$ can be rewritten as $(x^2y)(x^{-1}y^{-2})$, which is equal to $(x^{2-1})(y^{1-2})$. This reduces to xy^{-1}, or $\dfrac{x}{y}$. Notice that the result is the same as it would have been if we had simply subtracted the exponents of literal factors in the denominator from the exponents of the same literal factors in the numerator.

The algebraic rules for radicals also remain the same as those of arithmetic. In arithmetic, $\sqrt{4} = 4^{1/2} = 2$. Likewise, in algebra $\sqrt{a} = a^{1/2}$ and $\sqrt[n]{a} = a^{1/n}$.

MULTIPLYING MONOMIALS

If a monomial such as 3abc is to be multiplied by a numerical multiplier, for example 5, the coefficient alone is multiplied, as in the following example:

$$5 \times 3abc = 15abc$$

When the numerical factor is not the initial factor of the expression, as in x(2a), the result of the multiplication is not written as x2a. Instead, the numerical factor is interchanged with literal factors by use of the commutative law of multiplication. The literal factors are usually interchanged to place them in alphabetical order, and the final result is as follows:

$$x(2a) = 2ax$$

The rule for multiplication of monomials may be stated as follows: Multiply the numerical coefficients to form the coefficient of the product. Multiply the literal factors, combining exponents in like factors, to form the literal part of the product. The complete process is illustrated in the following example:

$$(2ab)(3a^2)(2b^3) = 12a^{1+2}b^{1+3}$$
$$= 12a^3b^4$$

Practice problems. Perform the indicated operations:

1. $(2x^2)(5x^5)$ 4. $(2^a)(2^b)$

2. $(-5ab^2)(2a^2b)$ 5. $(-4a^3)^2$

3. $(-4x^4y)(-3xy^4)$ 6. $(3a^2b)^2$

Answers:

1. $10x^7$ 4. 2^{a+b}

2. $-10a^3b^3$ 5. $16a^6$

3. $12x^5y^5$ 6. $9a^4b^2$

DIVIDING MONOMIALS

As may be expected, the process of dividing is the inverse of multiplying. Because 3 x 2a = 6a, 6a ÷ 3 = 2a, or 6a ÷ 2 = 3a. Thus, when the divisor is numerical, divide the coefficient of the dividend by the divisor.

When the divisor contains literal parts that are also in the dividend, cancellation may be performed as in arithmetic. For example, 6ab ÷ 3a may be written as follows:

$$\frac{(2)(3a)(b)}{3a}$$

265

Cancellation of the common literal factor, 3a, from the numerator and denominator leaves 2b as the answer for this division problem.

When the same literal factors appear in both the divisor and the dividend, but with different exponents, cancellation may still be used, as follows:

$$\frac{14a^3b^3x}{-21a^2b^5x} = \frac{(7)(2)a^2ab^3x}{(7)(-3)a^2b^3b^2x}$$

$$= \frac{2a}{-3b^2} \quad = -\frac{2a}{3b^2}$$

This same problem may be solved without thinking in terms of cancellation, by rewriting with negative exponents as follows:

$$\frac{14a^3b^3x}{-21a^2b^5x} = \frac{2a^{3-2}b^{3-5}x^{1-1}}{-3}$$

$$= \frac{2ab^{-2}}{-3} = \frac{2a}{-3b^2}$$

$$= -\frac{2a}{3b^2}$$

Practice problems. Perform the indicated operations:

1. $\dfrac{x^5}{x^6}$ 2. $\dfrac{a^9b^4}{a^6b^3}$

3. $\dfrac{a^2bc^2}{abc}$ 4. $\dfrac{a^2b}{ab^2}$

5. $\sqrt{16x^4y^6}$ 6. $\sqrt{x^{4a}y^{2a}}$

7. $\dfrac{5a^4b}{10a^2b^3}$ 8. $\dfrac{10x^2y^3z^4}{-5xy^2z^3}$

9. $\sqrt{100a^8b^4}$ 10. $\sqrt{a^6b^{6n}}$

Answers:

1. x^{-1}

2. a^3b

3. ac

4. $\dfrac{a}{b}$

5. $\pm\, 4x^2y^3$

6. $\pm\, x^{2a}y^a$

7. $\dfrac{a^2}{2b^2}$

8. $-\,2xyz$

9. $\pm\, 10a^4b^2$

10. $\pm\, a^3b^{3n}$

OPERATIONS WITH POLYNOMIALS

Adding and subtracting polynomials is simply the adding and subtracting of their like terms. There is a great similarity between the operations with polynomials and denominate numbers. Compare the following examples:

1. Add 5 qt and 1 pt to 3 qt and 2 pt.

$$
\begin{array}{r}
3\text{ qt} + 2\text{ pt} \\
5\text{ qt} + 1\text{ pt} \\
\hline
8\text{ qt} + 3\text{ pt}
\end{array}
$$

2. Add $5x + y$ to $3x + 2y$.

$$
\begin{array}{r}
3x + 2y \\
5x + \ \ y \\
\hline
8x + 3y
\end{array}
$$

One method of adding polynomials (shown in the above examples) is to place like terms in columns and to find the algebraic sum of the like terms. For example, to add 3a + b - 3c, 3b + c - d, and 2a + 4d, we would arrange the polynomials as follows:

$$
\begin{array}{r}
3a + b - 3c \\
3b + c - d \\
\underline{2a + 4d} \\
5a + 4b - 2c + 3d
\end{array}
$$

Subtraction may be performed by using the same arrangement—that is, by placing terms of the subtrahend under the like terms of the minuend and carrying out the subtraction with due regard for sign. Remember, in subtraction the signs of all the terms of the subtrahend must first be mentally changed and then the process completed as in addition. For example, subtract 10a + b from 8a - 2b, as follows:

$$
\begin{array}{r}
8a - 2b \\
\underline{10a + b} \\
-2a - 3b
\end{array}
$$

Again, note the similarity between this type of subtraction and the subtraction of denominate numbers.

Addition and subtraction of polynomials also can be indicated with the aid of symbols of grouping. The rule regarding changes of sign when removing parentheses preceded by a minus sign automatically takes care of subtraction. For example, to subtract 10a + b from 8a - 2b, we can use the following arrangement:

$$(8a - 2b) - (10a + b) = 8a - 2b - 10a - b$$
$$= -2a - 3b$$

Similarly, to add $-3x + 2y$ to $-4x - 5y$, we can write

$$(-3x + 2y) + (-4x - 5y) = -3x + 2y - 4x - 5y$$
$$= -7x - 3y$$

Practice problems. Add as indicated, in each of the following problems:

1. $3a + \ b$
   $\underline{2a + 5b}$

2. $(6s^3t + 3s^2t + st + 5) + (s^3t - 5)$

3. $4a + b + c$, $a + c - d$, and $3a + 2b + 2c$

4. $4x + 2y$
   $3x - \ y + z$
   $\underline{\ x \qquad - z}$

In problems 5 through 8, perform the indicated operations and combine like terms.

5. $(2a + b) - (3a + 5b)$

6. $(5x^3y + 3x^2y) - (x^3y)$

7. $(x + 6) + (3x + 7)$

8. $(4a^2 - b) - (2a^2 + b)$

Answers:

1. 5a + 6b   5. -(a + 4b)

2. $7s^3t + 3s^2t + st$ 6. $4x^3y + 3x^2y$

3. 8a + 3b + 4c - d   7. 4x + 13

4. 8x + y   8. $2(a^2 - b)$

MULTIPLICATION OF A POLYNOMIAL BY A MONOMIAL

We can explain the multiplication of a polynomial by a monomial by using an arithmetic example. Let it be required to multiply the binomial expression, 7 - 2, by 4. We may write this 4 x (7 - 2) or simply 4(7 - 2). Now 7 - 2 = 5. Therefore, 4(7 - 2) = 4(5) = 20. Now, let us solve the problem a different way. Instead of subtracting first and then multiplying, let us multiply each term of the expression by 4 and then subtract. Thus, 4(7 - 2) = (4 x 7) - (4 x 2) = 20. Both methods give the same result. The second method makes use of the distributive law of multiplication.

When there are literal parts in the expression to be multiplied, the first method cannot be used and the distributive method must be employed. This is illustrated in the following examples:

$$4(5 + a) = 20 + 4a$$

$$3(a + b) = 3a + 3b$$

$$ab(x + y - z) = abx + aby - abz$$

Thus, to multiply a polynomial by a monomial, multiply each term of the polynomial by the monomial.

Practice problems. Multiply as indicated:

1. $2a(a - b)$ 3. $-4x(-y - 3z)$

2. $4a^2(a^2 + 5a + 2)$ 4. $2a^3(a^2 - ab)$

Answers:

1. $2a^2 - 2ab$ 3. $4xy + 12xz$

2. $4a^4 + 20a^3 + 8a^2$ 4. $2a^5 - 2a^4b$

MULTIPLICATION OF A POLYNOMIAL BY A POLYNOMIAL

As with the monomial multiplier, we explain the multiplication of a polynomial by a polynomial by use of an arithmetic example. To multiply $(3 + 2)(6 - 4)$, we could do the operation within the parentheses first and then multiply, as follows:

$$(3 + 2)(6 - 4) = (5)(2) = 10$$

However, thinking of the quantity $(3 + 2)$ as one term, we can use the method described for a monomial multiplier. That is, we can multiply each term of the multiplicand by the multiplier, $(3 + 2)$, with the following result:

$$(3 + 2)(6 - 4) = [(3 + 2) \times 6 - (3 + 2) \times 4]$$

Now considering each of the two resulting products separately, we note that each is a binomial multiplied by a monomial.

The first is

$$(3 + 2)6 = (3 \times 6) + (2 \times 6)$$

and the second is

$$-(3 \times 2)4 = -[(3 \times 4) + (2 \times 4)]$$
$$= -(3 \times 4) - (2 \times 4)$$

Thus we have the following result:

$$(3 + 2)(6 - 4) = (3 \times 6) + (2 \times 6)$$
$$- (3 \times 4) - (2 \times 4)$$
$$= 18 + 12 - 12 - 8$$
$$= 10$$

The complete product is formed by multiplying each term of the multiplicand separately by each term of the multiplier and combining the results with due regard to signs.

Now let us apply this method in two examples involving literal numbers.

1. $(a + b)(m + n) = am + an + bm + bn$

2. $(2b + c)(r + s + 3t - u) = 2br + 2bs + 6bt - 2bu + cr + cs + 3ct - cu$

The rule governing these examples is stated as follows: The product of any two polynomials is found by multiplying each term of one by each term of the other and adding the results algebraically.

It is often convenient, especially when either of the expressions contains more than two terms, to place the polynomial with the fewer terms beneath the other polynomial and multiply term by term beginning at the left. Like terms of the partial products are placed one beneath the other to facilitate addition.

Suppose we wish to find the product of $3x^2 - 7x - 9$ and $2x - 3$. The procedure is

$$
\begin{array}{l}
3x^2 - 7x - 9 \\
2x - 3 \\
\hline
6x^3 - 14x^2 - 18x \\
 - 9x^2 + 21x + 27 \\
\hline
6x^3 - 23x^2 + 3x + 27
\end{array}
$$

Practice problems. In the following problems, multiply and combine like terms:

1. $(2a - 3)(a + 2)$

2. $(ax + b)(ax - b)$

3. $\dfrac{\begin{array}{l} x^3 + 5x^2 - x + 2 \\ 2x + 3 \end{array}}{}$

4. $\dfrac{\begin{array}{l} 2a^2 + 5ab - b^2 \\ a + b \end{array}}{}$

Answers:

1. $2a^2 + a - 6$

2. $a^2x^2 - b^2$

3. $2x^4 + 13x^3 + 13x^2 + x + 6$

4. $2a^3 + 7a^2b + 4ab^2 - b^3$

SPECIAL PRODUCTS

The products of certain binomials occur frequently. It is convenient to remember the form of these products so that they can be written immediately without performing the complete multiplication process. We present four such special products as follows, and then show how each is derived:

1. Product of the sum and difference of two numbers.

EXAMPLE: $(x - y)(x + y) = x^2 - y^2$

2. Square the sum of two numbers.

EXAMPLE: $(x + y)^2 = x^2 + 2xy + y^2$

3. Square of the difference of two numbers.

EXAMPLE: $(x - y)^2 = x^2 - 2xy + y^2$

4. Product of two binomials having a common term.

EXAMPLE: $(x + a)(x + b) = x^2 + (a + b)x + ab$

Product of Sum and Difference

The product of the sum and difference of two numbers is equal to the square of the first number minus the square of the second number. If, for example, x - y is multiplied by x + y, the middle terms cancel one another. The result is the square of x minus the square of y, as shown in the following illustration:

$$
\begin{array}{r}
x \;-\; y \\
x \;+\; y \\
\hline
x^2 \;-\; xy \\
+\; xy \;-\; y^2 \\
\hline
x^2 \qquad -\; y^2
\end{array}
$$

By keeping this rule in mind, the product of the sum and difference of two numbers can be written down immediately by writing the difference of the squares of the numbers. For example, consider the following three problems:

$$(x + 3)(x - 3) = x^2 - 3^2 = x^2 - 9$$

$$(5a + 2b)(5a - 2b) = (5a)^2 - (2b)^2 = 25a^2 - 4b^2$$

$$(7x + 4y)(7x - 4y) = 49x^2 - 16y^2$$

RATIONALIZING DENOMINATORS.— The product of the sum and difference of two numbers is useful in rationalizing a denominator that is a binomial. For example, in a fraction such as

$$\frac{2}{\sqrt{2} - 6}$$

the denominator can be altered so that no radical terms appear in it. (This process is called rationalizing.) The denominator must be multiplied by $\sqrt{2} + 6$, which is called the conjugate of $\sqrt{2} - 6$. Since the value of the original fraction would be changed if we multiplied only the denominator, our multiplier must be applied to both the numerator and the denominator. Multiplying the original fraction by

$$\frac{\sqrt{2} + 6}{\sqrt{2} + 6}$$

is, in effect, the same as multiplying it by 1.
The result of rationalizing the denominator of this fraction is as follows:

$$\frac{2}{\sqrt{2} - 6} \cdot \frac{\sqrt{2} + 6}{\sqrt{2} + 6} = \frac{2(\sqrt{2} + 6)}{(\sqrt{2})^2 - 6^2}$$

$$= \frac{2(\sqrt{2} + 6)}{2 - 36}$$

$$= \frac{2(\sqrt{2} + 6)}{2(1 - 18)}$$

$$= \frac{2(\sqrt{2} + 6)}{2(-17)}$$

$$= \frac{\sqrt{2} + 6}{-17}$$

MENTAL MULTIPLICATION.—The product of the sum and difference can be utilized to mentally multiply two numbers that differ from a multiple of 10 by the same amount, one greater and the other less. For example, 67 is 3 less than 70 while 73 is 3 more than 70. The product of 67 and 73 is then found as follows:

$$67(73) = (70 - 3)(70 + 3)$$

$$= 70^2 - 3^2 = 4,900 - 9 = 4,891$$

Square of Sum and Difference

The square of the SUM of two numbers is equal to the square of the first number plus twice the product of the numbers plus the square of the second number. The square of the DIFFERENCE of the same two numbers has the same form, except that the sign of the middle term is negative.

These results are evident from multiplication. When x and y represent the two numbers, we obtain

$$
\begin{array}{ll}
\begin{array}{r}
x + y \\
x + y \\
\hline
x^2 + xy \\
 + xy + y^2 \\
\hline
x^2 + 2xy + y^2
\end{array}
&
\begin{array}{r}
x - y \\
x - y \\
\hline
x^2 - xy \\
 - xy + y^2 \\
\hline
x^2 - 2xy + y^2
\end{array}
\end{array}
$$

Applying this rule to the squares of the binomials $3a + 2b$ and $3a - 2b$, we have the following two cases:

1. $(3a + 2b)^2 = (3a)^2 + 2(3a)(2b) + (2b)^2$

$$= 9a^2 + 12ab + 4b^2$$

2. $(3a - 2b)^2 = 9a^2 - 12ab + 4b^2$

The square of the sum or difference of two numbers is applicable to squaring a binomial that contains one or two irrational terms, as in the following examples:

1. $(\sqrt{3} + 8)^2 = (\sqrt{3})^2 + 2(8)(\sqrt{3}) + 64$

$$= 3 + 16\sqrt{3} + 64 = 67 + 16\sqrt{3}$$

2. $(\sqrt{3} - 8)^2 = (\sqrt{3})^2 - 2(8)(\sqrt{3}) + 64$

$$= 3 - 16\sqrt{3} + 64 = 67 - 16\sqrt{3}$$

3. $(\sqrt{5} + \sqrt{7})^2 = (\sqrt{5})^2 + 2\sqrt{5}\sqrt{7} + (\sqrt{7})^2$

$$= 5 + 2\sqrt{35} + 7 = 12 + 2\sqrt{35}$$

4. $(\sqrt{5} - \sqrt{7})^2 = 12 - 2\sqrt{35}$

The square of the sum or difference of two numbers can be applied to the process of mentally squaring certain numbers. For example, 82^2 can be expressed as $(80 + 2)^2$ while 67^2 can be expressed as $(70 - 3)^2$. We find that

$$(80 + 2)^2 = 80^2 + 2(80)(2) + 2^2$$

$$= 6,400 + 320 + 4 = 6,724$$

$$(70 - 3)^2 = 70^2 - 2(70)(3) + 3^2$$

$$= 4,900 - 420 + 9 = 4,489$$

Binomials Having a Common Term

The binomials $x + 2$ and $x - 3$ have a common term, x. They have two unlike terms, $+2$ and -3. The product of these binomials is

$$
\begin{array}{r}
x + 2 \\
x - 3 \\
\hline
x^2 + 2x \quad\quad \\
- 3x - 6 \\
\hline
x^2 - x - 6
\end{array}
$$

Inspection of this product shows that it is

278

obtained by squaring the common term, adding the sum of the unlike terms multiplied by the common term, and finally adding the product of the unlike terms.

Apply this rule to the product of 3y - 5 and 3y + 4. The common term is 3y; its square is $9y^2$. The sum of the unlike terms is -5 + 4 = -1; the sum of the unlike terms multiplied by the common term is -3y; and the product of the unlike terms is -5(4) = -20. The product of the two binomials is

$$(3y - 5)(3y + 4) = 9y^2 - 3y - 20$$

The product of two binomials having a common term is applicable to the multiplication of numbers like $\sqrt{3} + 7$ and $\sqrt{3} - 2$ which contain irrational terms. For example,

$$(\sqrt{3} + 7)(\sqrt{3} - 2) = (\sqrt{3})^2 + 5\sqrt{3} - 14$$
$$= 3 + 5\sqrt{3} - 14$$
$$= -11 + 5\sqrt{3}$$

Practice problems. In problems 1 through 4, multiply and combine terms. In 5 through 8, simplify by using special products.

1. $(x + 4)(x + 2)$

2. $(\sqrt{a} - b)^2$

3. $(7a + 4b)(7a - 4b)$

4. $(ax + y)^2$

5. $\dfrac{2}{\sqrt{2} - 2}$

6. 48(52)

7. $(\sqrt{3} + 7)^2$

8. $(73)^2$

Answers:

1. $x^2 + 6x + 8$

2. $a - 2b \sqrt{a} + b^2$

3. $49a^2 - 16b^2$

4. $a^2x^2 + 2axy + y^2$

5. $-(\sqrt{2} + 2)$

6. $(50 - 2)(50 + 2)$
   $= 2496$

7. $52 + 14 \sqrt{3}$

8. $(70 + 3)(70 + 3)$
   $= 5329$

DIVISION OF A POLYNOMIAL BY A MONOMIAL

Division, like multiplication, may be distributive. Consider, for example, the problem $(4 + 6 - 2) \div 2$, which may be solved by adding the numbers within the parentheses and then dividing the total by 2. Thus,

$$\frac{4 + 6 - 2}{2} = \frac{8}{2} = 4$$

Now notice that the problem may also be solved distributively.

$$\frac{4 + 6 - 2}{2} = \frac{4}{2} + \frac{6}{2} - \frac{2}{2}$$
$$= 2 + 3 - 1$$
$$= 4$$

CAUTION: Do not confuse problems of the type just described with another type which is similar in appearance but not in final result. For example, in a problem such as $2 \div (4 + 6 - 2)$ the beginner is tempted to divide 2 successively by 4, then 6, and then -2, as follows:

$$\frac{2}{4 + 6 - 2} \neq \frac{2}{4} + \frac{2}{6} - \frac{2}{2}$$

Notice that we have canceled the "equals" sign, because 2 ÷ 8 is obviously not equal to 1/2 + 2/6 - 1. The distributive method applies only in those cases in which several different numerators are to be used with the same denominator

When literal numbers are present in an expression, the distributive method must be used, as in the following two problems:

$$1. \ \frac{2ax + aby + a}{a} = \frac{2ax}{a} + \frac{aby}{a} + \frac{a}{a}$$

$$= 2x + by + 1$$

$$2. \ \frac{18ab^2 - 12bc}{6b} = \frac{18ab^2}{6b} - \frac{12bc}{6b}$$

$$= 3ab - 2c$$

Quite often this division may be done mentally, and the intermediate steps need not be written out.

DIVISION OF A POLYNOMIAL BY A POLYNOMIAL

Division of one polynomial by another proceeds as follows:

1. Arrange both the dividend and the divisor in either descending or ascending powers of the

same letter.

2. Divide the first term of the dividend by the first term of the divisor and write the result as the first term of the quotient.

3. Multiply the complete divisor by the quotient just obtained, write the terms of the product under the like terms of the dividend, and subtract this expression from the dividend.

4. Consider the remainder as a new dividend and repeat steps 1, 2, and 3.

EXAMPLE:

$$(10x^3 - 7x^2y - 16xy^2 + 12y^3) \div (5x - 6y)$$

SOLUTION:

$$
\require{enclose}
\begin{array}{r}
2x^2 + xy - 2y^2 \\
5x - 6y \enclose{longdiv}{10x^3 - 7x^2y - 16xy^2 + 12y^3} \\
\underline{10x^3 - 12x^2y} \\
5x^2y - 16xy^2 \\
\underline{5x^2y - 6xy^2} \\
-10xy^2 + 12y^3 \\
\underline{-10xy^2 + 12y^3} \\
\end{array}
$$

In the example just shown, we began by dividing the first term, $10x^3$, of the dividend by the first term, $5x$, of the divisor. The result is $2x^2$. This is the first term of the quotient.

Next, we multiply the divisor by $2x^2$ and subtract this product from the dividend. Use the remainder as a new dividend. Get the second term, xy, in the quotient by dividing the first term, $5x^2y$, of the new dividend by the first term, 5x, of the divisor. Multiply the divisor by xy and again subtract from the dividend.

Continue the process until the remainder is zero or is of a degree lower than the divisor. In the example being considered, the remainder is zero (indicated by the double line at the bottom). The quotient is $2x^2 + xy - 2y^2$.

The following long division problem is an example in which a remainder is produced:

$$
\begin{array}{r}
x^2 - x + 3 \\
x + 3 \overline{\smash{\big)}\ x^3 + 2x^2 + 5} \\
\underline{x^3 + 3x^2} \\
-x^2 \\
\underline{-x^2 - 3x} \\
3x + 5 \\
\underline{3x + 9} \\
-4
\end{array}
$$

The remainder is -4.

Notice that the term -3x in the second step of this problem is subtracted from zero, since there is no term containing x in the dividend. When writing down a dividend for long division, leave spaces for missing terms which may enter during the long division process.

In arithmetic, division problems are often arranged as follows, in order to emphasize the relationship between the remainder and the divisor:

$$\frac{5}{2} = 2 + \frac{1}{2}$$

This same type of arrangement is used in algebra. For example, in the problem just shown, the results could be written as follows:

$$\frac{x^3 + 2x^2 + 5}{x + 3} = x^2 - x + 3 - \frac{4}{x + 3}$$

Remember, before dividing polynomials arrange the terms in the dividend and divisor according to either descending or ascending powers of one of the literal numbers. When only one literal number occurs, the terms are usually arranged in order of descending powers.

For example, in the polynomial $2x^2 + 4x^3 + 5 - 7x$ the highest power among the literal terms is x^3. If the terms are arranged according to descending powers of x, the term in x^3 should appear first. The x^3 term should be followed by the x^2 term, the x term, and finally the constant term. The polynomial arranged according to descending powers of x is $4x^3 + 2x^2 - 7x + 5$.

Suppose that $4ab + b^2 + 15a^2$ is to be divided by $3a + 2b$. Since 3a can be divided evenly into $15a^2$, arrange the terms according to descending powers of a. The dividend takes the form

$$15a^2 + 4ab + b^2$$

Synthetic Division

Synthetic division is a shorthand method of dividing a polynomial by a binomial of the form $x - a$. For example, if $3x^4 + 2x^3 + 2x^2 - x - 6$ is to be divided by $x - 1$, the long form would be as follows:

$$
\begin{array}{r}
3x^3 + 5x^2 + 7x + 6 \\
x - 1 \overline{\smash{\big)}\ 3x^4 + 2x^3 + 2x^2 - x - 6} \\
\underline{3x^4 - 3x^3} \\
+ 5x^3 + 2x^2 \\
\underline{+ 5x^3 - 5x^2} \\
+ 7x^2 - x \\
\underline{+ 7x^2 - 7x} \\
+ 6x - 6 \\
\underline{+ 6x - 6}
\end{array}
$$

Notice that every alternate line of work in this example contains a term which duplicates the one above it. Furthermore, when the subtraction is completed in each step, these duplicated terms cancel each other and thus have no effect on the final result. Another unnecessary duplication results when terms from the dividend are brought down and rewritten prior to subtraction. By omitting these duplications, the work may be condensed as follows:

$$
\begin{array}{r}
3x^3 \ +5x^2 \ +7x \quad +6 \\
x - 1 \overline{\smash{\big)}\ 3x^4 \ +2x^3 \ +2x^2 \ -x \quad -6} \\
\underline{-3x^3 \ -5x^2 \ -7x \quad -6} \\
+5x^3 \ +7x^2 \ +6x \quad 0
\end{array}
$$

The coefficients of the dividend and the constant term of the divisor determine the results of each successive step of multiplication and subtraction. Therefore, we may condense still further by writing only the nonliteral factors, as follows:

$$
\begin{array}{r}
3 \quad +5 \quad +7 \quad +6 \\
- 1 \overline{\smash{\big)}\ 3 \quad +2 \quad +2 \quad -1 \quad -6} \\
\underline{-3 \quad -5 \quad -7 \quad -6} \\
3 \quad +5 \quad +7 \quad +6 \quad 0
\end{array}
$$

Notice that if the coefficient of the first term in the dividend is brought down to the last line, then the numbers in the last line are the same as the coefficients of the terms in the quotient. Thus we do not really need to write a separate line of coefficients to represent the quotient. Instead, we bring down the first coefficient of the dividend and make the subtraction "subtotals" serve as coefficients for the rest of the quotient, as follows:

$$
\begin{array}{r|rrrrr}
x - 1 & 3 & 2 & 2 & -1 & -6 \\
 & & -3 & -5 & -7 & -6 \\
\hline
 & 3 & 5 & 7 & 6 & 0
\end{array}
$$

The unnecessary writing of plus signs is also eliminated here.

The use of synthetic division is limited to divisors of the form $x - a$, in which the degree of x is 1. Thus the degree of each term in the quotient is 1 less than the degree of the corresponding term in the dividend. The quotient in this example is as follows:

$$3x^3 + 5x^2 + 7x + 6$$